18th International Conference on Mechanical, Materials and Manufacturing Engineering (ICMMME 2016)

London, United Kingdom
18 - 19 January 2016

ISBN: 978-1-5108-4662-3

Printed from e-media with permission by:

Curran Associates, Inc.
57 Morehouse Lane
Red Hook, NY 12571

proceedings.com

Some format issues inherent in the e-media version may also appear in this print version.

World Academy of Science, Engineering, and Technology (WASET)
http://www.waset.org

Printed by Curran Associates, Inc. (2017)

Additional copies of this publication are available from:

Curran Associates, Inc.
57 Morehouse Lane
Red Hook, NY 12571 USA
Phone: 845-758-0400
Fax: 845-758-2633
Email: curran@proceedings.com
Web: www.proceedings.com

TABLE OF CONTENTS

Electrochemical Performance of Al-Mn_2O_3 Based Electrode Materials

Noor Ul Ain Bhatti, M. Junaid Khan, Javed Ahmad, Murtaza Saleem, Shahid M. Ramay, Saadat A. Siddiqi

***Abstract*—**Manganese oxide is being recently used as electrode material for rechargeable batteries. In this study, Al incorporated Mn_2O_3 compositions were synthesized to study the effect of Al doping on electrochemical performance of host material. Structural studies were carried out using X-ray diffraction analysis to confirm the phase stability and explore the lattice parameters, crystallite size, lattice strain, density and cell volume. Morphology and composition were analyzed using field emission scanning electron microscope and energy dispersive X-ray spectroscopy, respectively. Dynamic light scattering analysis was performed to observe the average particle size of the compositions. FTIR measurements exhibit the O-Al-O and O-Mn-O and Al-O bonding and with increasing the concentration of Al, the vibrational peaks of Mn-O become sharper. An enhanced electrochemical performance was observed in compositions with higher Al content.

***Keywords*—**Mn_2O_3, electrode materials, energy storage and conversion, electrochemical performance.

I. Introduction

THE tailoring of layered oxide materials have been widely discussed subject as a source for portable energy storage devices. Numerous materials have been proposed that proved to be a good candidate for advanced technologies including batteries, fuel cells, solar cells and super capacitors [1], [2]. The advancement in search of energy storage materials enforce us to make cost efficient and maximum storage capacity materials which could be used for portable electronic devices and hybrid electronic vehicles. In recent electronic applications, lithium ion batteries are preferably used as they have high coulombic energy efficiency, magnificent low self-discharge rate and lack of memory effect [3], [4]. Graphite is enormously used as commercial material for lithium ion batteries but due to its low storage capacity, energy, power density, longer life cycle and improved safety there is a need to search new materials as layered oxides as a use for better electrode material [5]. To enhance storage capacity of Li ion batteries, transition metal oxides in nano-structured materials have been extensively studied [6], [7]. Among all transition metal oxides [8] such as manganese oxide [9], [10], ruthenium oxide [11], cobalt oxide [12], [13], nickel oxide [14], copper oxide [15], manganese oxide is considered as the most reliable electrode material because of their abundance, non-toxicity, low cost, multiple high theoretical capacitance and environmental compatibility. Ruthenium oxide is toxic and other materials have also limited in their use so we prefer manganese oxide [16], [17]. A number of research studies have been done on synthesis of nanostructured manganese oxide with different experimental techniques such as physical vapor deposition, chemical bath deposition, sputtering [18] and sol-gel methods [19]. Elemental doping of simple metal oxides to form mixed metal oxides has been a way to boost up their electrochemical activities, electronic conductivity and to minimize their cost efficiency. In this case, binary systems doping is preferable such as NiMn, MnCo [20], MnFe [21], NiCo [22], SnAl oxides and ternary metal oxides systems like Mn-Ni-Co [23], [24], Co-Ni-Cu [25] and Mn-Ni-Cu oxides [26]. In present research work, we study sol-gel synthesized nanostructured Mn_2O_3 with doping combination of Al. The samples are physically characterized by X-Ray diffraction (XRD), field emission scanning electron microscope (FESEM), energy dispersive X-ray spectroscopy (EDS), dynamic light scattering, Fourier transform infrared spectroscopy (FTIR), and electrochemical measurements. The effect of Al substitution on all these properties of Mn_2O_3 was precisely studied and found that electrochemical performance was significantly enhanced.

Noor Ul Ain Bhatti, M. Junaid Khan, and Javed Ahmad are with the Laboratory of Experimental and Theoretical Physics, Department of Physics, Bahauddin Zakariya University (60800), Multan, Pakistan.

Murtaza Saleem is with the Department of Physics, Syed Babar Ali School of Sciences and Engineering (SBASSE), Lahore University of Management Sciences (LUMS), Opposite Sector U, DHA, Lahore 54792, Pakistan.

Shahid M. Ramay is with the College of Science, Department of Physics and Astronomy, King Saud University, Riyadh, Saudi Arabia.

Saadat A. Siddiqi is with the Interdisciplinary Research Centre in Biomedical Materials, COMSATS Institute of Information Technology, M.A. Jinnah Campus, Lahore-54600, Pakistan (Corresponding Author; phone: +92-300-4866616; e-mail: saadat.anwar@ciitlahore.edu.pk).

II. Experimental Section

A. Synthesis

Al doped Mn_2O_3 nanostructured compositions were synthesized using well established sol-gel based fuel agent assisted self-combustion technique. Stoichiometric molar ratios of analytical research grade aluminium nitrate [$Al(NO_3)_2.9H_2O$] and manganese nitrate [$Mn(NO_3)_2.4H_2O$] were use as the sources of Al and Mn, respectively. In this preparation, glycine was used as the fuel agent. Calculated molar amounts of reagents for undoped and Al doped Mn_2O_3 were weighed on a precise digital balance and subsequently dissolved in 25 ml distilled water. The solution was kept on hot place at 250 °C with constant stirring inside a clean fume hood environment. The temperature of the attained gel was raised to 350 °C and stirring was subsequently stopped. Fine powder of brownish black color was obtained after intense exothermic combustion reaction. The resulting products were dried and calcinated at 600 °C for 4 hours, subsequently, a

International Science Index, Chemical and Molecular Engineering Vol:11, No:1, 2017 waset.org/Publication/10006152 dark brownish powder was obtained for each composition. The powder was grinded well using mortar and pestle until fine, homogenous and uniform sized particles were achieved.

B. Characterizations

Structural studies were carried out using XRD instrument (Bruker D-2 Phaser) operating at 30 kV and 10 mA using Cu K-alpha radiation (λ=1.5406 Å) and a step scan size of 0.05°. A FESEM (FEI Nova 450) was used to observe the morphology of the samples. Qualitative and quantitative elemental compositions were confirmed using EDS spectrometry (Oxford Instrument Inca X-Act). FTIR analysis were carried out using Bruker Alpha instrument. Chinstruments 660E was used to extract the first charge-discharge curves and cyclic performance of all the synthesized compositions.

III. Results and Discussion

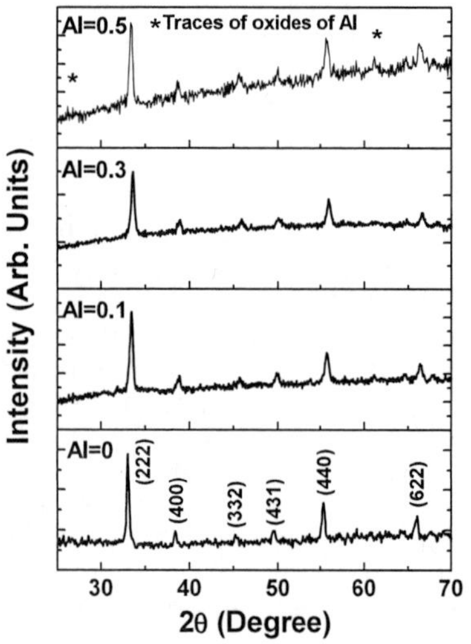

Fig. 1 XRD spectra of undoped and Al doped compositions

The XRD spectra of undoped and Al doped Mn_2O_3 are shown in Fig. 1. All the peaks in XRD patterns are exactly matched with the host Mn_2O_3 standard pattern containing JCPD card number 76-0150. However, some traces of oxides of Al were observed and clearly appeared in composition with highest content of Al and this showing some different behavior as compared to other spectra. It is clear that there are visible peak shifts to higher angle in the spectrum which is consistent to the previously reported substitutions [27], [28] with lower ionic radii. Crystallite size, cell volume, lattice constant and lattice strain were calculated and observed a significant variation in it with doping of Al content. Crystallite size and cell volume was observed as 43-53 nm and 784-824 $Å^3$, respectively as shown in Table I. A significant variation in lattice constant (9.22-9.38 Å) and lattice strain (2.64-15.57) was also observed revealed the doping of Al content in to Mn_2O_3 structure.

The microstructural studies were carried out using FESEM equipped with EDX detector. Fig. 2 presents the FESEM micrographs of undoped and Al doped Mn_2O_3 samples. It is visible through images that morphologies are oriented in non-uniform arrangement but they have distinct boundaries and clear shapes like microspheres and fillets with some compact shapes, as well as few cavities in them. Particles are present in random sizes which are estimated as 94.57 nm for $Al_{0.1}Mn_{1.9}O_3$, 144 nm for $Al_{0.3}Mn_{1.7}O_3$ and 172 nm for $Al_{0.5}Mn_{1.5}O_3$ samples. The particles in Al doped Mn_2O_3 are observed to be arranged more densely because of relatively smaller size of Al.

Fig. 3 represents the EDX spectra for all of the four samples. The EDX analysis reveals that Mn, O and Al contents are present in the approximate stoichiometric ratios in the samples, which ensure the incorporation of Al constituents in the host MnO_2 structure. Traces of very small carbon content might visible just because of the sample remnant. Quantitative EDX analysis is approximately according to the dissolved molar ratio for added elements in all samples as given in Table II.

FITR spectra of Mn_2O_3 and doped with different concentrations of Al are studied in the range 400-550 cm^{-1} as shown in Fig. 4 (a) which reveals the presence of O-Mn-O and O-Al-O stretching modes. The transmittance spectra of undoped and doped samples have considerable differences among them. Most of the transmission bands are stretched between the wavelength 400-550 cm^{-1}. The wide band appearing from 500 cm^{-1} onward correspond to vibrational frequencies of coordinate O-Al-O bond [29]. Low frequency band in the viscinity of 419 cm^{-1} correspond to the change in O-Mn-O bond angle [30]. Other bonds in the range 400-500 cm^{-1} is correspond to the Al-O vibrations [31]. As we increase Al concentration, we observe that Mn-O vibrational peak at ~ 500 cm^{-1} appear to be sharper.

TABLE I
Lattice Constant, Crystallite Size, Cell Volume, Lattice Strain and Density of Compositions

Sample	Lattice Constants (Å)	Crystallite Size (nm)	Cell Volume ($Å^3$)	Lattice strain($\times10^{-3}$)	Density (gcm^{-1})
Undoped	9.38	44	824.50	2.64	5.08
$Al_{0.1}Mn_{1.9}O_3$	9.27	53	795.82	12.04	6.17
$Al_{0.3}Mn_{1.7}O_3$	9.22	53	784.7	15.57	6.26
$Al_{0.5}Mn_{1.5}O_3$	9.27	45	796.3	11.75	6.17

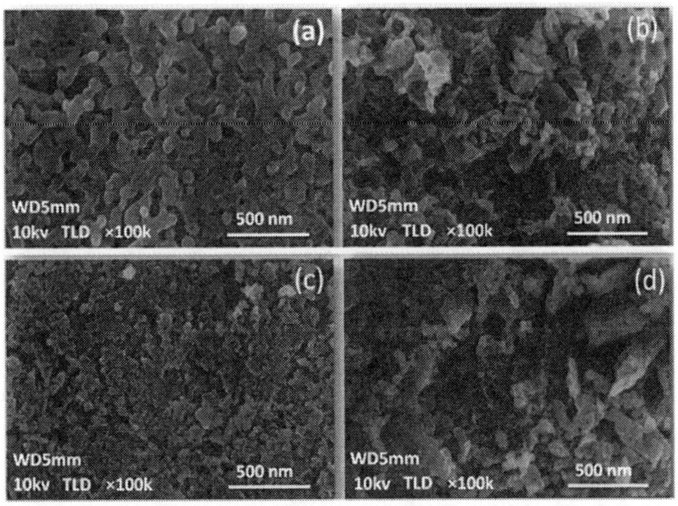

Fig. 2 FESEM micrographs of undoped and Al doped compositions

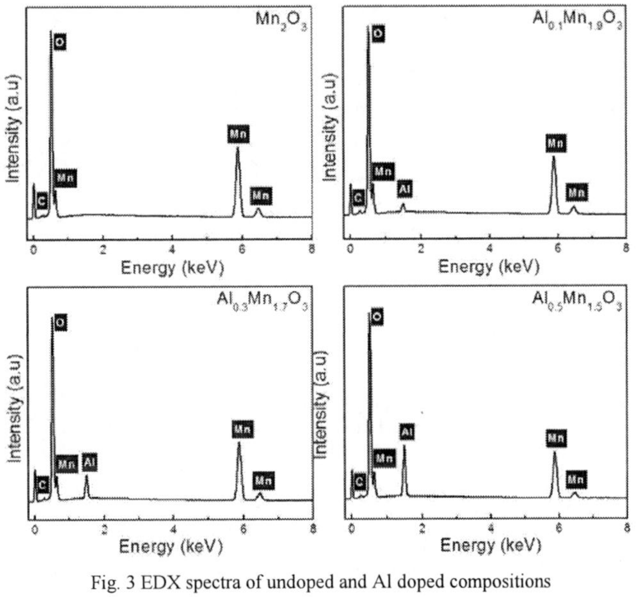

Fig. 3 EDX spectra of undoped and Al doped compositions

TABLE II
EDX Elemental Compositional Measurements

	Undoped		$Al_{0.1}Mn_{1.9}O_3$		$Al_{0.3}Mn_{1.7}O_3$		$Al_{0.5}Mn_{1.5}O_3$	
	At.%	Wt.%	At. %	Wt.%	At. %	Wt.%	At.%	Wt.%
O	62.27	34.08	64.15	37.53	63.56	37	64.28	40.69
Al	0	0	1.69	1.66	4.32	4.24	8.78	9.38
Mn	34.34	64.53	29.17	58.61	28.64	57.24	21.86	47.52
C	3.39	1.39	4.99	2.2	3.48	1.51	5.08	2.41
Total				100				

It is well understood that shape and structure of materials have direct influence on the electrochemical properties in respect of cyclability and performance [32]. However, there are yet few works reported to enhance the electrochemical performance with additional doping in Mn_2O_3 like materials. The electrochemical performance of all $Al_xMn_{2-x}O_3$ compositions was obtained to determine the Al doping effect. Figs. 4 (b), (c) show the first cycle curves for charging and discharging process of all the samples. A wide sharp plateau can be seen in all first discharge curves at various values. The similar fact was also observed in some previous reports [33]. The obtained initial discharge capacity for Al substituted compositions is higher than that of un-doped Mn_2O_3. The lowest charge capacity was observed for Mn_2O_3 while highest for composition doped with maximum value of Al. Cycling performance of all compositions was measured as shown in Fig. 4 (d). Saturation in capacity was reached after a sharp initial decrease. It was found that Al doped compositions show enhanced capacity values after maximum number of cycles. It can be realized that Al substitution in Mn_2O_3 improves the electrochemical properties.

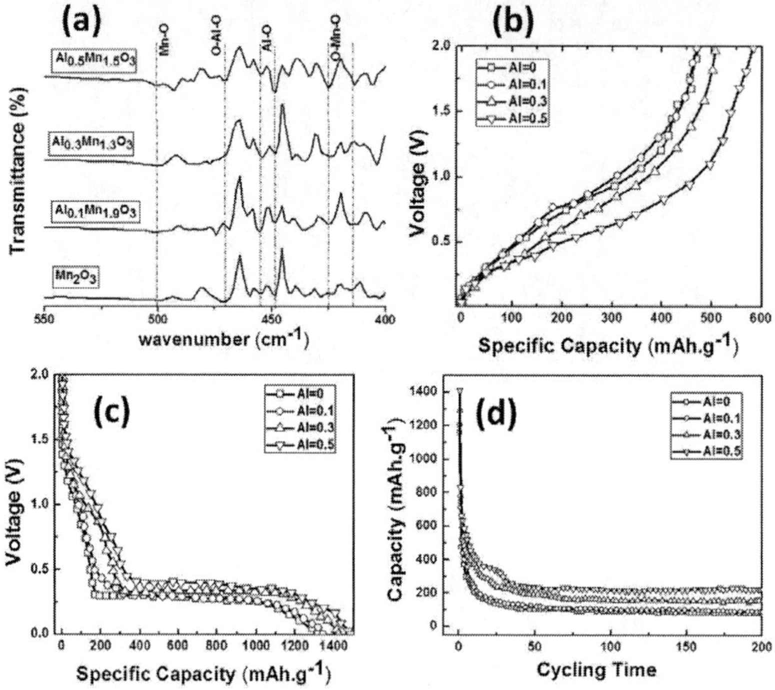

Fig. 4 (a) FTIR spectra and (b-d) electrochemical performance of undoped and Al doped compositions

IV. Conclusions

Al incorporated Mn_2O_3 compositions were successfully synthesized to study the effect of Al doping on electrochemical performance of host material. XRD analysis confirm the phase stability and explore the variation in lattice parameters, crystallite size, lattice strain, density and cell volume. Morphology and elemental composition analysis were in good agreement with the dissolved concentrations of Al in host material. The FTIR results exhibit the O-Al-O and O-Mn-O and Al-O bonding and with increasing the concentration of Al, the vibrational peaks become sharper. An indication of enhanced electrochemical performance in compositions with incorporation of Al content is considerably important for application point of view.

Acknowledgement

Supported by Deanship of Scientific Research at King Saud University through the research group project No. RG 1435–004.

References

[1] C. Liu, F. Li, L.P. Ma, H.M. Cheng, Adv. Mater. 22 (2010) E28-E62.
[2] M.S. Whittingham, MRS Bull. 33 (2008) 411-419.
[3] B. Scrosati, J. Hassoun, Y.K. Sun, Energy Environ. Sci. 4 (2011) 3287-3295.
[4] M. Broussely, P. Biensan, B. Simon, Electrochim. Acta. 45 (1999) 3-22.
[5] P. Kristin, A.S. Vijay, H.H. Laurence, H. Yoyo, S.M. Ying, V.V. Anton, S. Venkat, K. Robert, C. Gerbrand, J. Phys. Chem. Lett. 1 (2010) 1176-1180.
[6] H.B. Wu, J.S. Chen, H.H. Hng, X.W. Lou, Nanoscale 4 (2012) 2526-2542.
[7] A.C. Dillon, L.A. Riley, Y.H. Kim, C. Ban, D.T. Gillaspie, A. Pesaran, S.H. Lee, Nanostructured Metal Oxide Anodes, NREL, DOE Merit Review (2009).
[8] A.A. Salvatore, P. Bruce, B. Scrosati, J.M. Tarascon, W.V. Schalkwijk, Nat. Mater. 4 (2005) 366-377.
[9] D. Yang, J. Power Sources 228 (2013) 89-96.
[10] H. Zhimi, X. Xiao, C. Chen, T. Li, L. Huang, C. Zhang, J. Su, L. Miao, J. Jiang, Y. Zhang, J. Zhou, Nano Energy 11 (2015) 226-234.
[11] L.D. Burke, J.F. Healy, J. Electroanal. Chem. 124 (1981) 327-332.
[12] J.E. Arthur, J.M. Meredith, N.R. Philip, T.B. Alexis, T.D. Tilley, J. Phys. Chem. C 113 (2009) 15068-15072.
[13] J.B. Wu, Y. Lin, X.H. Xia, J.Y. Su, Q.Y. Shi, Electrochim. Acta 56 (2011) 7163-7170.
[14] X.H. Huang, J.P. Tu, B. Zhang, C.Q. Zhang, Y. Li, Y.F. Yuan, H.M. Wu, J. Power sources 161 (2006) 541-544.
[15] J.Y. Xiang, J.P. Tu, X.H. Huang, Y.Z. Yang, J. Solid State Electrochem. 12 (2008) 941-945.
[16] F. Li, Y.X. Zhang, M. Huang, Y. Xing, L.L. Zhang, Electrochim. Acta 154 (2015) 329- 337.
[17] Z.K. Ghouri, M.S. Akhtar, A. Zahoor, A.A. Barakat, W. Han, M. Park, B. Pant, P.S. Saud, C.H. Lee, H.Y. Kim, J. Alloys compd 642 (2015) 210-215.
[18] A.A. Salvatore, P. Bruce, B. Scrosati, J.M. Tarascon, W.V. Schalkwijk, Nat. Mater. 4 (2015) 366-377.
[19] N.R. Ravinder, G.R. Ramana, J. Power Sources 124 (2003) 330-337.
[20] J Kuei, W.C. Hseieh, W.T. Tsai, J. Alloys Compd. 461 (2008) 667-674.

[21] M.T. Lee, J.K. Chang, Y.T. Hsieh, W.T. Tsai, J. Power Sources, 185 (2008) 1550-1556.
[22] D.P. Dubal, A.D. Jagadale, S.V. Patil, C.D. Lokhande, MRS Bull. 47 (2012) 1239-1245.
[23] J.M. Luo, B. Gao, X Ghang, MRS Bull. 43 (2008) 1119-1125.
[24] C.S. Ferreria, R.R. Passos, L.A. Pocrifka, J. Power Sources 271 (2014) 104-107.
[25] T.C. Wen, H.M. Kang, Electrochim. Acta 43 (1998) 1729-1745.
[26] D.L. Fang, Z.D. Chen, B.C. Wu, Y. Yan, C.H. Zheng, Mater. Chem. Phys. 128 (2011) 311- 316.
[27] G. Adylov, V. Voronov, L. Sigalov, Inorg. Mater. 23 (1988) 1867-1870.
[28] R.V. Sagar, S. Buddhudu, Spectrochim. Acta A 75 (2010) 1218-1222.
[29] N. Varghese, M. Hariharan, D.A.B. Cherian, D.P.V. Sreenivasan, J. Paul, A.K.A. Antony, Int. J. Sci. Res. Pub. 4 (2014) 2250-3153.
[30] K. Lipping, M. Zhang, Z.H. Liu, K. Ooi, Spectrochim. Acta A 67 (2007) 864-869.
[31] S.A. Hosseini, A. Niaei, D. Salari, Open J. Phys. Chem. 1 (2011) 23-27.
[32] R. Ma, M. Wang, P. Tao, Y. Wang, C. Cao, G. Shan, S. Yang, L. Xi, J.C.Y. Chung, Z. Lu, J. Mate. Chem. A 1 (2013) 15060-15067.
[33] L. Hu, Y. Sun, F. Zhang, Q Chen, J. Alloys Compd. 576 (2013) 86-92.

Automated User Story Driven Approach for Web-Based Functional Testing

Mahawish Masud, Muhammad Iqbal, M. U. Khan, Farooque Azam

***Abstract*—**Manual writing of test cases from functional requirements is a time-consuming task. Such test cases are not only difficult to write but are also challenging to maintain. Test cases can be drawn from the functional requirements that are expressed in natural language. However, manual test case generation is inefficient and subject to errors. In this paper, we have presented a systematic procedure that could automatically derive test cases from user stories. The user stories are specified in a restricted natural language using a well-defined template. We have also presented a detailed methodology for writing our test ready user stories. Our tool "Test-o-Matic" automatically generates the test cases by processing the restricted user stories. The generated test cases are executed by using open source Selenium IDE. We evaluate our approach on a case study, which is an open source web based application. Effectiveness of our approach is evaluated by seeding faults in the open source case study using known mutation operators. Results show that the test case generation from restricted user stories is a viable approach for automated testing of web applications

***Keywords*—**Automated testing, natural language, user story modeling, software engineering, software testing, test case specification, transformation and automation, user story, web application testing.

I. Introduction

A remarkable spread of web applications into the areas of commerce and communications has made web applications a significant and fairly larger part of the software industry [1]. Web applications and services will change the landscape of IT architecture, by providing a new means of service delivery for enterprises. As predicted by Hagel and Brown [2], companies are outsourcing or purchasing their services over the internet. The economic significance of web applications also increases the importance of controlling and improving the quality of web applications [3]-[5]. Web applications therefore require through and systematic test approaches for testing.

A common approach for testing web applications is by manually writing record-and-replay test cases and executing them in a test execution tool, such as Selenium or HPE Unified Functional Testing. Manual system level testing of web applications from requirements can be a significant challenge. Manually writing test cases from requirements is not only time consuming, but also difficult to maintain when the requirements change [6]. Moreover, the tester has to write the test cases manually.

When testing is done manually then the tester has to write the test cases covering most or all of the features of the system depending on the available resources, such as time and manpower. Moreover, instead of following a thorough testing procedure, the tester might consider exploring the user interface of the application and cover as many features as possible. The success of such a manual testing approach largely depends on the expertise, domain knowledge, and enthusiasm of the tester [7]. A lack of these factors might lead to poor testing of the application. As the software evolves, it becomes a difficult task to maintain the traceability of requirements to test cases. Some requirement changes may result in test cases becoming obsolete and because of the lack of traceability, identifying such test cases is labor intensive.

Our aim in this paper is to automate the test case generation process from requirements expressed through user stories in natural language. The intension to employ user stories for test generation is twofold: (i) It is a prevalent practice in industry for gathering the functional requirements of web based applications from user's point of view (ii) User stories are written in natural language, which makes it easier for the user to express, authenticate and prioritize their requirements. It is focused on the result that the user wants to achieve, and in the language that he/she understands. This helps to understand user requirements clearly and in short span of time. Moreover, the user stories outline functionality that is required from a system without getting into lengthy details. The user may like to accept the user story or delay it for development at a later stage, without wasting much time on gathering lengthy and detailed requirement's first. By postponing details for the later stage, the user stories can help well in understanding of problem domain.

The presented approach of test generation from user stories utilizes the requirements written in user stories. The user stories are to be written as per our suggested template. The template uses a restricted language that allows the user stories to be processed for test generation. The approach has been applied on an open source case study, Moodle, which is a course management system written in PHP. The evaluation has been done by using known mutation operators of HTML and PHP. Results show that the approach was able to detect all the 195 seeded equivalent non-equivalent mutants.

Even though the testable user stories may seem to be tightly coupled to the graphical interface, this is not a major obstacle in its practical application. A common practice in web application testing is to write test cases in record-and-replay tools that are coupled with graphical interface (GUI). In case of a change in GUI, the entire test cases (potentially in hundreds) relating to the change have to be manually

Mahawish Masud is a Student at CE&ME, National University of Sciences and Technology (NUST). Pakistan (e-mail: mahawish.masud@nust.edu.pk)

Dr. Farooque Azam, is a Professor at CE&ME, NUST.

modified. In our case, the modification is simpler as the user only needs to change a few user stories rather than the test cases. Overall, the contributions of this paper are: (i) we suggest an approach for writing user stories in a format that allows automated test generation; (ii) we have developed a test generation strategy to obtain test sequences and test data from the user stories; (iii) the approach is successfully applied on an open source case study.

The rest of the paper has been organized as: Section II explains the related work. Section III explains the proposed approach. Section IV explains evaluation of our technique, and in Section V, we conclude our work.

II. Related Work

There are works that derive test cases from requirements expressed in natural language, by employing natural language processing (NLP) techniques, but most of these techniques do not fully support automatic test case generation [8]; for instance, Zhang et al. [7] and Sarmiento et al. [9] generate the test cases from natural language without test data. Escalona et al. [10] can generate test cases but need manual intrusion for providing input values to the generated test cases. Nevertheless, there are techniques available in literature [6] that have used NLP for test case automation and also support the test data generation, but they use the domain models and OCL constraints to generate the test data and invariably help in the test case generation process. All these approaches are contrary to our approach in the sense that we want to fully automate the test case generation from requirements expressed in natural language and through user stories. We also tend to avoid dealing with the domain models. Hametner et al. [11] and Carvalho et al. [12] have also generated the test cases from requirements with the help of restricted grammar and dictionary, which is quite similar to our approach too, but the dictionary used in their approach is subject to change for every project, and the restricted grammar is not able to fully express the requirements of various systems. Unlike their approach, we tend to use our approach flexible enough to be used on different kinds of systems.

Literature suggests that user stories are a much better way of gathering user requirements [13]-[15]. For web applications, writing requirements through user stories have shown to be much more practical [13], [16] since user stories are written in natural language and simple enough to be understandable by both the user and the domain experts. To investigate, if literature suggests any approaches that have used the user stories for the test case automation, it was found out that Cucumber testing frame work [16]-[18] suggests writing the user stories in a natural language; however, it involves writing the user stories using a particular language called “Gherkin”. However, “Gherkin” is a very inflexible camouflaged form of Ruby, which apparently allows user to write user story in a syntax which is close enough to English language, but it is not English [19], [20]. Such a technique is not feasible in our case, because primarily it does not support the automatic test case generation right from the requirements and secondly, and we want to keep the user stories simple and flexible enough to be understandable by both user and domain experts.

To summarize, various techniques are available in literature that do support automated test case generation from requirements that are expressed in natural language. However, issue with almost all of the techniques is that either they require certain level of manual intervention for their correction and completion or they involve complex models. Such approaches are contrary to our aim, which is to automatic test case generation from requirements given in natural language through user stories.

III. User Story Based Test Generation Approach

Our approach mainly focuses on derivation of automatic test cases from testable user stories. The approach consists of two parts: (i) A user story driven modeling approach that deals with how to write the user story and (ii) An automated approach for test generation. Our approach shall encompass training steps:

A. User Story Modeling Approach

In this section, we will go over the important features of user story modeling approach for the automated test generation. The approach consists of a user story template, and a set of restriction rules that can be used to improve the understanding and reduce the ambiguity in the use of natural language. Next we have defined a keyword list of data input types that are commonly used to input data in many web based applications. The attributes of these data input types give the user the flexibility to pick and choose the test data. We have also defined action type that would translate the actions on keywords, and invariably help in the process of automation. We have also illustrated how a user story could be written using our user story template, restriction rules, keywords, and action types.

This section discusses various techniques used in the literature for generation of automatic test cases from requirements.

1) Template for Writing User Stories

The proposed template for writing the user stories allows the writer to express the user story in restricted natural language. The template includes basic information about the user story, e.g. action type, data input types, and attributes of data input types that can be used for test generation. The idea is to keep the user story template simple enough to be understandable by both the users and the domain experts. The fields in the user story template are pretty simple to comprehend and are part of many templates. The Conditions of Satisfaction (CoS) do not contain any branching or conditional flows. We recommend separate user story for each branching flow.

The CoS should be written using our pre-defined data input types and action keywords, which shall help in the automation of the user story. Moreover, the user stories have to be written keeping in view the set of grammar rules that should not be violated.

TABLE I
USER STORY TEMPLATE

User Story Name	User Story name, and it generally starts with a verb.	
Pre-conditions	What should hold true before the user story is executed.	
CoS	Also, called "Happy Path" and specifies the steps that constitute the main successful path.	
	Steps (1….n)	List of events in the main successful path.
	Post-condition	What should hold true after the CoS terminate

2) Writing Testable User Stories

In order to better understand how a user story should be written, using our user story template, let us consider an example: Consider a user who intends to make a new account on Moodle which is an open source course management system. The user would have to follow a few steps to make the new account. Ideally, he/she would have to go to the sign-up page, fill in the required information, and hit 'create my new account' button. Now, this user story when written as per our prescribed format is depicted in Table II.

TABLE II
USER STORY OF MOODLE WRITTEN AS PER USER STORY TEMPLATE

Story Name	Moodle Signup
Description	To test the signup functionality of Moodle
Actors	User
Pre conditions	User has a working internet connection.
Conditions of Satisfaction	
1. The user *goes to* the **PAGE**:/r.php	
2. The user *enters* user name in the **TEXTFIELD_T** : id_username, Admin	
3. The user *enters* password in the **TEXTFIELD_T**: id_password, Admin@22	
4. The user *enters* email in the **TEXTFIELD_email**: id_email, administrator@hotmail.com	
5. The user *enters* email in the **TEXTFIELD_email**: id_email2, administrator@hotmail.com	
6. The user *enters* first name in the **TEXTFIELD_T**: id_firstname, Amy	
7. The user *enters* last name in the **TEXTFIELD_T**: id_lastname, Smith	
8. The user *enters* city in the **TEXTFIELD_T**: id_city, ISB	
9. The user *selects* country in the **DROPDOWN_N**: id_country,Pakistan	
10. The user *clicks* the **BUTTON**: id_submitbutton	
11. The user *sees* the **TEXT**: Welcome Admin	
end_story	

The user story steps have to be written keeping in mind three things (i) no restriction rules are violated; (ii) keyword list; (iii) action type. Note that the user story contains number of keywords specified to the approach that we will discuss.

3) Restriction Rules

We adopt the natural language restriction rules proposed by Yue et al. for writing user stories [21]. The rules have been well defined to remove the ambiguities of the natural language and invariably help in improving the structure of the sentences in user stories and bring clarity to sentences. The rules have been divided into not-allowed keyword (Nak) rules and sentence structure rules. The user story steps are written keeping in line the restriction rules. Note that due to space issues we enlisted only few rules. In order to understand how these Nak rules are used in user story, let us consider the first Nak, which defines that IF/AND/BUT are not allowed in user story steps. Now, in order to comprehend this, consider our user story steps in example expressed in Table II. The Step 1 and Step 2 of the user story cannot be joined in a single step, using "IF", "AND", and "BUT". If this rule is violated and two steps are merged into one, then it would be difficult to parse the user story steps. The intention is to bring clarity to the use of natural language, and at the same time get a clear and precise meaning of the user action.

TABLE III
LIST OF NOT ALLOWED KEYWORDS (NAK 1-NAK 11)

#	Not Allowed Keywords(Nak)	Explanation
Nak 1	No IF/AND/BUT allowed in user stories	Compound sentences are difficult to parse
Nak 2	No use of UNLESS/ EXCEPT	
Nak 3	Do not use model verbs. For instance, MIGHT/POSSIBLY/COULD	Model verbs as well as adverbs introduce uncertainty.
Nak 4	Adverbs should not be used. (e.g. very)	
Nak 5	Negative adverbs are not allowed for instance NEVER and HARDLY. However, it is permissible to use no or not.	
Nak 6	The sentence should be simple, should not contain any conjunctions. One action per sentence is allowed.	It is hard to predict the correct sequence of more than one action in a sentence.

TABLE IV
LIST OF SENTENCE STRUCTURE RULES (R1-R6)

	Description	Explanation
R1.	Action steps should be sequential.	Implements use of our restricted user story template and the interaction types.
R2.	Only present tense is allowed.	Describes what the system does rather than what the system has done or will do.
R3.	Make use of active voice rather than passive voice	
R4.	The subject of the sentence in CoS should be either a system or an actor.	Explicitly defining sender and receiver would remove ambiguity in NL.
R5.	No Actor to Actor interactions are allowed	
R6.	The interactions between the actors and the systems should be clearly defined, without omitting the sender and receiver	

The sentence structure rules help to improve the structure of the sentences, and bring clarity to natural language. Table IV lists all sentence structure rules from R1 to R6.

The sentence structure rules are pretty easy to comprehend; for instance, rule 1 of sentence structure rules states that action steps should be sequential. Refer to Table VIII, each user story step contains only one action keyword and all the user story steps are written in a sequential order.

4) Keywords List of Data Input Types and their Definitions

We have proposed a list of keywords of commonly used data input types that are used in web-based applications to input certain type of data. Then, list can be extended further; however, we have defined a total of 14 keywords that define the most commonly used data input types and are outlined in Table V.

TABLE V
KEYWORD LIST AND THEIR DESCRIPTIONS

Sr.	Data Input Types	Description
1.	TextField_T	Use for combinations of text and numbers. 0 character minimum 255 characters maximum.
2.	TextField_M	Memo: It is used for bigger amounts of text. 0 character minimum 65,536 characters maximum.
3.	Page / URL/ Hyperlink	Contains links to other web pages or files.
4.	email	["[a-zA-Z0-9\\.]+ @ [a-zA-Z0-9\\-_\\.]+\\ . [a-zA-Z0-9]{3}"]' '
5.	date	Format DD:MM:YY
6.	time	Format: HH:MM:SS
7.	Year	2-digit or 4-digit format allowed. Allowed values in 4-digit format: 1800 - 2100. Allowed values in 2-digit format: 60 - 69, representing years from 1960 – 2069
8.	Image	Maximum 500MB. Format JPEG, TIFF, GIF, BMP, PNG
9.	currency	15 digits of whole numbers, and 4 decimal places. Define the country currency.
10.	phone number	Contains a combination of numeric digits (0-9)
11.	Postal code	5-digit number from (0-9)
12.	Boolean	Either TRUE or FALSE
13.	Id number	13-digit number (0-9)
14.	Calendar	A combination of numeric digits (0-9) and characters (A-Z) 9-character long

The 14 data input types as defined in Table IV have been elaborated further. For instance, the first data input type TEXTFIELD_T is explained in Table VI. Due to space limitation, rest of the date input types is not appended here.

The data input type TEXTFIELD_T is used to input small amounts of texts in a textfield; for instance, when a user intends to write a "user name" in some website, he/she may choose TEXTFIELD_T as the data input type. Next he/she can choose from the attributes of the data input type TEXTFIELD_T. For instance, the user can then choose that the maximum no. of characters that he/she allows is 8, and only upper case letters from A-Z are allowed and numerical values from 1-10 are allowed. So, the user name will contain values within the defined attributes of the data type TEXTFIELD_T.

TABLE VI
DATA INPUT TYPE TEXTFIELD_T

Data Type: TEXTFIELD (TEXT)		
1.	Format:	0 character minimum 255 characters maximum.
2.	Type:	**TEXTFIELD _T** (for TEXT)
3.	Max no. of characters	26
4.	Null values allowed	Yes/No
5.	Upper case(A-Z) allowed	Yes/No
6.	Lower case (a-z) allowed	Yes/No
7.	Numerical Values Allowed	Yes/No
8.	Alphanumeric_characters_allowed	a-z, 0-9, Punctuation*, Symbols*

**punctuations:* [] () {} * , : = ; ... # ' " " < << > >> ! ? _ - '

**symbols:* & * \ @ • ^ ¿ + – ' " ‴ § ¡ ¶ ~ ÷

With the purpose of better understanding how the keywords and their attributes are used while writing the user story, refer to step 2 in Table II. The keyword used in step 2 is TEXTFIELD_T, which indicates that the user intends to input some data in the text field, like in the said case the user wants to input the user name. Now, once the keyword TEXTFIELD_T has been defined by the user, the next step is to define the attributes of the keyword selected, for instance the attributes related to keyword TEXTFIELD_T as outlined in Table VI. The user may like to allow only upper case alphabets to be used for user name, or he/she may allow a combination of upper and lower case alphabets with numerical values. So, for each keyword used, the attributes of the keywords have to be selected.

5) Action Type

The steps in user stories are written using the predefined action types, which are enlisted in Table VII.

TABLE VII
LIST OF ACTION TYPE

	Action Type	Keywords
1.	Goes	PAGE
2.	Enters	TEXTFIELD_T TEXTFIELD_M TEXTFIELD_EMAIL TEXTFIELD_DATE TEXTFIELD_YEAR
3.	Clicks	Button Link
4.	Selects	DROPDOWN_T RADIOBUTTON_T CHECKBOX_T
5.	Sees	TEXT

B. Automated Approach for Test Generation

These action types define the intended user actions and are used in conjunction with the keywords. For instance, the action type "goes" suggests that the user wants to go to a page (URL). So, the action keyword "goes" is used with the keyword PAGE. The step in a user story then could be like "The user goes to the PAGE". Similarly, action type "enter" shall be used along with a keyword that is used to input data; for instance, it shall be used with data input type "TEXTFIELD _T" or with data input type "date" or "email".

The generation of test case from user story involves three steps: (i) writing and validating the user story as per the prescribed format (ii) approach for mapping from user story to test sequence (iii) approach for test data generation.

1) Proposed Strategy for Writing and Validating the User Story

Given the user story has been written as per our prescribed format. It will then be checked for any limitations. First, it will be ensured that user story is in the required format.

Each first level field name which is also called label shall be read first. It shall be ensured that user has entered data against all the labels. The data entered by user along with the label is called a step.

Each step will be sequentially read. When the label has CoS, then second level fields, i.e. CoS steps will be read. For each CoS.step, the algorithm will check three things. First, it will check if there is any violation of restriction rules, like if

International Science Index, Computer and Information Engineering Vol:11, No:1, 2017 waset.org/Publication/10006309

some Naks have been used or Sentence Structure Rules have been violated. Next, it will check if only defined keywords from keywords list have been used. Lastly, it will check if it contains only defined action types. If the three conditions are not met, an error will be generated else the keyword name and action type will be saved in CoSStep Result.

2) Approach for Mapping User Story to Test Sequence Illustrated.

To convert the user story into a test case, three things have to be recorded from the user story (i) Test Sequence (ii) Test data (iii) Action type.

Both test sequence and test data shall be recorded from the given example in a manner identified at Table VIII.

TABLE VIII
EXTRACTION OF TEST SEQUENCE AND TEST DATA FROM USER STORY

Sr.	Test Sequence		Test Data
	Data Input Types	**Target**	
1.	PAGE	signup.php	
2.	TEXTFIELD_T	id_username	Admin
3.	TEXTFIELD_T	id_password	Admin@22
4.	TEXTFIELD_email	id_email	administrator@hotmail.com
5.	TEXTFIELD_email	id_email2	administrator@hotmail.com
6.	TEXTFIELD_T	id_firstname	Amy
7.	TEXTFIELD_T	id_lastname	Smith
8.	TEXTFIELD_T	id_city	ISB
9.	DROPDOWN_N	id_country	Pakistan
10.	BUTTON	id_submitbutton	
11.	TEXT		Module Testing Page for Thesis

The "test sequence" and "test data" extracted from the given user story will be mapped to the Selenium commands. The mapping is explained in Table X. Note that the approach is specific to Selenium.

TABLE IX
MAPPING TO SELENIUM COMMANDS

	Test Sequence			Test Data
	Data Input Types	**Selenium Command**	**Target**	**Value**
1.	**PAGE**	Open	/login/index.php	
2.	**Textfield_T**	Type	id= uname	Admin
3.	**Textfield_T**	Type	id=upass	Admin@22
4.	**BUTTON**	ClickandWait	id=loginbtn	
5.	**TEXT**	VerifyTextPresent		Module Testing Page for Thesis

The test data can be generated randomly by generating values among the allowed attributes of the data input type. For instance, if the user intends to input a user name in textfield_t. The user can choose that the user name is 5-character long with no numerical value and only alphabets in upper and lower case are allowed. So, the test data can be generated within the allowed parameters defined by the user.

Action types are defined to express what action has to be applied on the data input types. In order to better understand the Action type lets first extract the action type from the given example and enlist them in Table X.

TABLE X
ACTION KEYWORD MAPPING

	Action Keyword	Data Input Type	Required Parameters
1.	Goes	PAGE	URL
2.	Enters	TEXTFIELD_T TEXTFIELD_M TEXTFIELD_EMAIL TEXTFIELD_DATE TEXTFIELD_YEAR	Element ID through: Element Name Element Class Name Element Tag Name Element CSS Selector Element Partial Link Element Xpath
3.	Clicks	Button Link	Button ID Button Name Button Class Name Button Tag Name Button Link Text Button CSS Selector Button Partial Link Button Xpath
4.	Selects	DROPDOWN_T RADIOBUTTON_T CHECKBOX_T	Element ID
5.	Sees	TEXT	Text

The action keyword "goes to" will be mapped to the data input type PAGE, which expects a URL of the page. Similarly, the action keyword enters is mapped to data input type TEXTFIELD_T, TEXTFILED_M, and expects an element id, which could preferably be the id of the element, but could be done through link text, name, partial link, tag name or CSS select.

3) Approach for Mapping from User Story to Test Sequence

The algorithm shall explain the mapping from user story to test sequence. It reads the steps in user story under the label "condition of satisfaction", and extracts the action type. The action type is then mapped to Selenium commands. Also, the format of allowed data input types in each step are recorded and saved.

Algorithm of Mapping from User Story to Test Sequence

Require: *CoSStepResult*, (step number in Condition of satisfaction, Action name, Keyword, Keyword format)
Require: *AcK*, List of Action types e.g. Clicks/Types/Open
Require: *Format*, List of Action type
Ensure *CoSStepResult*, *List of pairs* <ActionName, AllowedDataType, and Allowed Data Type Format>

```
// Read CoSStep from CoSStepResult list
//Mapping the Action type from CoS
// Command is the name of Selenium command. Action types are mapped to Selenium commands.
If CoSStep has a AcK then
      Action. Name ← AcKName
      ActionKeywordMapping(Action.Name)
      CoSStep ← Action.Name U Command
      CoSStepResult ← CoSStep
      CoSStep.next
end if
If CoSStep has a ADT
      ADT. Name ← ADTName
      AllowedDataTypes(ADT.Name)
      CoSStep ← ADT.Name U Format
```

```
        CoSStepResult ← CoSStep
        CoSStep.next
    end if
```

4) Approach for Test Data Generation

An important aspect of software testing is the generation of test data, which is the process of creating dataset for checking the adequacy of web based applications. The test data could be either be actual or artificial test data.

The test data can be generated using the attributes of the allowed data input types. The user first selects the data input type for instance the user has selected textfiled_T as the allowed data input type. The user then defines the attributes of that data input type, for instance the user may choose that only 26 characters are allowed, with a lower-case letters and numerical values between 1-5. Now the data can be generated by varying combinations of data attributes that have been allowed by the user. We have generated random test data.

5) Tool for Writing User Story

A tool "Test-o-Matic" has been developed which can automate the testing of web applications using user story. It takes a user story as an input written in a specified format and converts it into an executable test case to run with the help of a Selenium. Selenium web driver helps to automate the interaction with the web application.

The tool has a built-in text editor which aids the user in writing the user story or multiple user stories to build up a test suite. It also extends the facility to open up a pre-saved user story in the editor in .txt format. To write a user story the user has to identify the identification factors of the web elements, which could easily be located using the Webpage inspector within the tool.

The user story must contain the action, data type, id to identify, web elements and value to pass. Fig. 1 elaborates a user story written in Text editor of the tool. The action type is highlighted in green, the input data types are highlighted pink and web element id's are on the right hand column under the 'Identifiers List'. The test data are on the extreme right of every line. For instance, the test data in line 2, Fig. 1 is 'admin'. The user story can be written in the text editor or a pre-save user story in .txt format can also be opened.

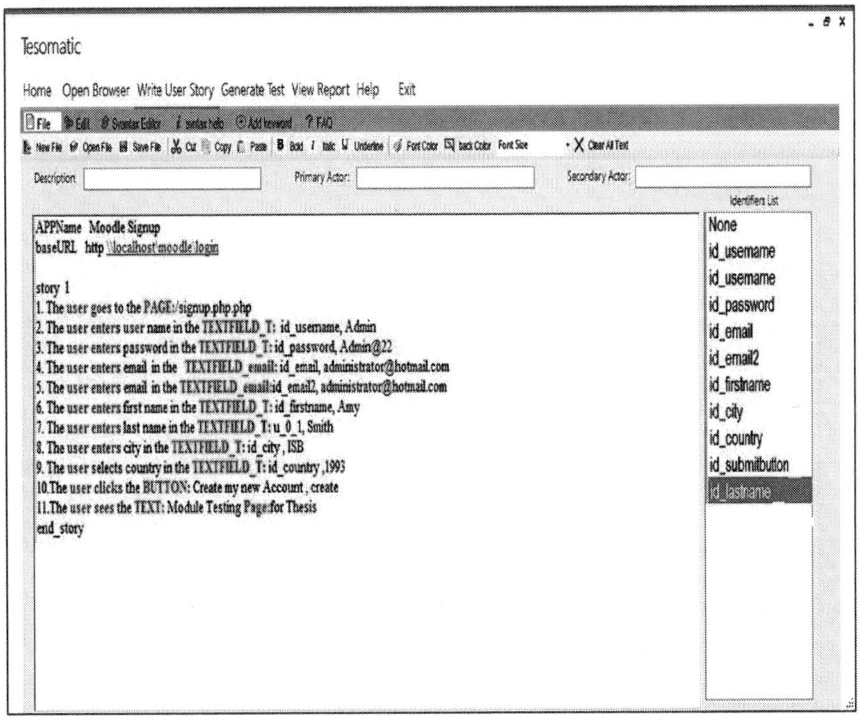

Fig. 1 Writing a user Story in Test-o-Matic tool

Once a user has written and saved the user story, the "generate test" automatically generates the test cases and redirects to a browser and runs the test case.

IV. Evaluation

We evaluate the presented approach by applying it on web based application called "Moodle", which is an open source course management system written in PHP. We selected this application as it is a widely-used system and is of sufficient complexity. We conduct mutation analysis by seeding X faults and evaluating the fault detection effectiveness of our proposed approach.

For the experiment, we wrote a total of 12 user stories to document the requirements of Moodle. A total of 33 test cases were then written against the corresponding user stories.

For mutation analysis, we seeded mutants corresponding to mutation operators of both HTML and PHP. The HTML mutation operators are taken from Upsorn et al. [22], whereas the PHP mutation operators are taken from Mutagenesis testing framework [23]. We seeded a total of 79 mutants corresponding to HTML mutation operators and 98 mutants corresponding to the PHP mutation operators. The HTML mutation operators are outlined in Table XI, whereas the PHP mutation operators used in the analysis are highlighted in Table XII.

TABLE XI
HTML MUTATION OPERATORS

	Mutation Operator Abbreviation	Mutation Operator
1	WLR	Simple Link Replacement
2	WLD	Simple Link Deletion
3	WFR	Form Link Replacement
4	WTR	Transfer Mode Replacement
5	WHR	Hidden Form Field Replacement
6	WHD	Hidden Form Field Deletion
7	WIR	Server-Side Include Replacement
8	WID	Server-Side Include Deletion

TABLE XII
PHP MUTATION OPERATORS

	Mutation Operator Abbreviation	Mutation Operator
1	Add	Addition
2	CNE	Conditional Negation Equal
3	Equal	Equal
4	False	False Value
5	GT	Greater Than
6	Identical	Identical
7	LT	Less Than
8	LLA	Logical Lower And
9	LLO	Logical Lower Or
10	LN	Logical Not
11	LO	Logical Or
12	NE	Not Equal
13	NI	Not Identical

Refer to Table XII, the mutation operator Web Link Replacement (WLR), replaces destination of a simple link transition specified in <a> tag with another destination, Simple Link Deletion (WLD), removes the destination link specified in <a> tag, Form Link Replacement (WFR), changes destination of a form link transition to another destination in the same domain of the targeted web applications, Transfer Mode Replacement (WTR), replaces all POST requests to GET and vice versa, Hidden Form Field Replacement (WHR), alters the value attributes of <input> tag to a space, null, zero or an empty string, Hidden Form Field deletion (WHD), removes an entire block of tag < input> of type hidden, Server-Side include replacement (WIR), changes file attributes of include directives to another destination in the same domain of the targeted web applications, server-side include deletion(WID), removes an entire include directive from the html file.

The results of application of our proposed approach are shown in Table XIII for HTML mutants and in Table XIV for PHP mutants. For HTML mutants, our approach was able to kill all the seeded mutants. For PHP mutants, the test cases generated corresponding to our approach was able to kill 96 out of 97 mutants, whereas the remaining one live mutant had no effect on the behavior of the application and was therefore classified as an equivalent mutant.

TABLE XIII
SUMMARY OF MUTANTS AND RESULTS

Html File	Mutants (HTML)								Total	Killed	Score (%)
	WLR	WLD	WFR	WTR	WHR	WHD	WIR	WID			
Index form	2	5	3	4	6	2	4	2	28	26	92
index	-	1	-	-	-	-	-	2	3	3	100
Forgot password form	1	-	-	-	-	-	1	1	3	3	100
Forgot password	7	-	-	-	-	-	5	2	14	14	100
logout	3	-	-	-	-	-	3		6	4	66
signup	-	-	-	-	-	-	2		2	2	100
Signup form	2	1	-	-	-	-	3		6	6	100
Change Password	10	1	-	-	-	-	4	3	18	16	89
Change Password form	1	-	-	-	-	-	4	1	6	3	50
Set password form	-	-	-	-	-	-	2	-	2	2	100
	-	-	-	-	-	-	-	-	86	79	92

The mutation operator Addition, replaces '+' with '-', Conditional Negation-Equal(CNE), replaces '==' to '!=', Equal, replaces, FalseValue, replaces FALSE with TRUE, Greater Than, replaces '>' with '>=', Identical, replaces '===' with '!==', LessThan, replaces '<' with '>=', LogicalLowerAnd, replaces 'and' with 'or', LogicalLowerOr, replaces 'or' with 'and', LogicalNot, replaces '!' with a blank string LogicalOr, replaces '||' with'&&', NotEqual, replaces '!= or <>' with '==' and NotIdentical, replaces '!==' with '==='. A total of 97 mutants were hand seeded into the application, out of which 96 were killed, and 1 was live mutant, and 1 was equivalent mutatnt. The live mutants were not killed because it was dealing with user session data, which was not dealt in our test cases. Same could be dealt in future for better results. One equivalent mutant was ignored, yielding to a total of 97 mutants. Table XIV outlines the results.

V. CONCLUSION

Web Applications form a large portion of the overall software industry. These applications range from simple discussion based application to highly critical financial applications. An essential element for precise functioning of these websites require testing, which should be thorough as well and systematic.

TABLE XIV
SUMMARY OF MUTANTS AND RESULTS

	Mutants (php)													Total	Killed	Score (%)
	Add	CNE	Equal	False	GT	Identical	LT	LLA	LLO	LN	LO	NE	NI			
Index form	-	5	2	-	-	-	-	-	-	7	1	-	-	15	14	93
index	1	4	2	1	1	2	2	3	1	17	-	-	6	40	40	100
Forgot pwd form	-	-	-	-	1	-	-	1	-	4	-	-	-	6	6	100
Forgot Pwd	-	-	-	-	-	-	-	3	1	2	-	-	-	6	6	100
logout	-	-	-	-	-	-	-	-	-	2	-	-	-	2	2	100
signup	-	-	-	-	-	-	-	1	-	3	-	-	-	4	4	100
Signup form	-	-	-	-	-	-	-	-	-	1	-	-	3	4	4	100
Change Pwd	-	-	-	-	1	1	-	-	-	5	-	-	-	7	7	100
Change Pwd form	-	1	1	-	-	-	-	-	-	5	-	-	1	8	7	87
Set Pwd form	-	-	-	-	1	-	-	-	-	3	-	-	1	5	5	100
Total	-	11	5	1	4	3	2	8	2	49	1	1	4	97	96	98

A common approach for testing such applications is by either writing or use a record and replay test tool which could run test cases corresponding to the user stories. Manual writing of such test cases is not scalable, especially because of the frequent changes in applications. We have presented an approach that could test web based applications automatically from restricted user stories. The restricted user stories can be written using the restricted natural language, following some set rules. The rules remove the ambiguity in natural language and improve sentence structure. A well-defined restricted user story template has been defined which allows the tester to write the restricted user story. The process is simplified using our Restricted User Story (RUST) tool 'Test-o-Matic" which enables a tester to write the user stories in restricted user story format using the restricted grammar. It also automatically checks if the restriction rules have been followed or not. The tool then automatically generates the test cases by processing the developed restricted user stories. The tool is integrated with a well-known open source record-and-replay tool, Selenium. The fault detection effectiveness of the presented approach is evaluated on an open source course management system (Moodle) using mutation analysis. The approach was tested on open source web-based application by seeding faults into the source code and the results prove that test case generation from restricted user stories can be an effective technique for automated web based application testing.

REFERENCES

[1] Offutt, J., Quality attributes of web software applications. IEEE software, 2002. 19(2): p. 25.
[2] Hagel, J. and J.S. Brown, Your next IT strategy. Harvard business review, 2001. 79(9): p. 105-115.
[3] Kreger, H., Web services conceptual architecture (WSCA 1.0). IBM Software Group, 2001. 5: p. 6-7.
[4] Zo, H., D.L. Nazareth, and H.K. Jain. Measuring reliability of applications composed of web services. in System Sciences, 2007. HICSS 2007. 40th Annual Hawaii International Conference on. 2007. IEEE.
[5] Ricca, F. and P. Tonella. Analysis and testing of web applications. in Proceedings of the 23rd international conference on Software engineering. 2001. IEEE Computer Society.
[6] Wang, C., et al. Automatic generation of system test cases from use case specifications. in Proceedings of the 2015 International Symposium on Software Testing and Analysis. 2015. ACM.
[7] Zhang, M., et al. A systematic approach to automatically derive test cases from use cases specified in restricted natural languages. in International Conference on System Analysis and Modeling. 2014. Springer.
[8] Sinha, A., S.M. Sutton Jr, and A. Paradkar. Text2Test: Automated inspection of natural language use cases. in 2010 Third International Conference on Software Testing, Verification and Validation. 2010. IEEE.
[9] Sarmiento, E., J.C.S. do Prado Leite, and E. Almentero. C&L: Generating model based test cases from natural language requirements descriptions. in Requirements Engineering and Testing (RET), 2014 IEEE 1st International Workshop on. 2014. IEEE.
[10] Escalona, M., et al., An overview on test generation from functional requirements. Journal of Systems and Software, 2011. 84(8): p. 1379-1393.
[11] Hametner, R., D. Winkler, and A. Zoitl. Agile testing concepts based on keyword-driven testing for industrial automation systems. in IECON 2012-38th Annual Conference on IEEE Industrial Electronics Society. 2012. IEEE.
[12] Carvalho, G., et al. Test case generation from natural language requirements based on SCR specifications. in Proceedings of the 28th Annual ACM Symposium on Applied Computing. 2013. ACM.
[13] Lucassen, G., et al. Forging high-quality user stories: towards a discipline for agile requirements. in 2015 IEEE 23rd international requirements engineering conference (RE). 2015. IEEE.
[14] North, D., Introducing bdd. Better Software, March, 2006.
[15] North, D., What's in a story. 2007, February.
[16] Saravana. K.M, G.N.B., Rajkumar, Dr. A. Kovalan, " Case Study On Agile User Stories Prioritization Using Imaginative Standard". IJERA Vol. 2(no. 5): p. pp. 472-480.
[17] Cucumber, https://cucumber.io/. Retrieved June, 2015.
[18] JBehave, http://jbehave.org/. Retrieved July, 2015.
[19] Testing, C., http://www.jackkinsella.ie/2011/09/26/why-bother-with-cucumber-testing.html. Access Date, March 2016.
[20] Drawbacks, C., https://www.jimmycuadra.com/posts/please-don-t-use-cucumber/. Access Date, March 2016.
[21] Yue, T., L.C. Briand, and Y. Labiche. A use case modeling approach to facilitate the transition towards analysis models: Concepts and empirical evaluation. in International Conference on Model Driven Engineering Languages and Systems. 2009. Springer.
[22] Praphamontripong, U. and J. Offutt. Applying Mutation Testing to Web Applications. in ICST Workshops. 2010.
[23] Framework, P.M.t., A PHP 5.3+ Mutation Testing framework. https://github.com/padraic/mutagenesis, January 2016.

Resources and Strategies towards the Development of a Sustainable Construction Materials Industry in Botswana

G. Malumbela, E. U. Masuku

Abstract**—**The economy of Botswana has increased extensively since its independence. In contrast to this increase, the construction industry which is one of the key indicators of a developing nation continues to be highly dependent on imported building material products from the neighbouring countries of South Africa, Namibia, Zimbabwe, and Zambia. Only two companies in the country currently blend cement. Even then, the overwhelming majority of raw materials used in the blends are imported. Furthermore, there are no glass manufacturers in Botswana. The ceramic industry is limited to the manufacture of clay bricks notwithstanding a few studios on crockery and sanitary ware which nonetheless use imported clay. This paper presents natural resources and industrial waste products in Botswana that can be used for the development of sustainable building materials. It also investigates at the distribution and cost of other widely used building materials in the country. Finally, the present paper looks at projects and national strategies aimed at a country-wide development of a sustainable building materials industry together with their successes and hitches.

Keywords**—**Botswana construction industry, construction materials, natural resources, sustainable materials.

I. Introduction

BOTSWANA is one of the fastest growing economies in the world with a nominal GDP per capita that grew by about eight times in the past twenty years from BWP 8800.8 million in 1995 to BWP 66795.9 million in 2015 [1]. The construction industry that drives this GDP is however, heavily reliant on imported ready-made building materials. Fig. 1 shows the import bill for cement, glass and ceramic products from 2010 to 2015 [2]-[4]. As shown in the figure, in the past five years the import bill of blended cement has ranged from BWP 566 million to BWP 628 million, of glass and glassware has ranged from BWP 140.6 million to BWP 189.2 million annually, whilst that for ceramic products ranged from BWP 170 million to BWP 220 million. In 2015, the combined import bill for these materials was BWP 942 million which is about 1.4% of the GDP. Note that these are conventional materials used worldwide, and their production has now become standardised.

The approach of making materials available to the industry through import of ready-made conventional materials could be the least challenging since they do not need any processing. It is however certainly not sustainable and offers little contribution to the general welfare of the nation which already exhibits a continued increase in unemployment rate and poverty incidences [5], [6]. It is therefore necessary to look at local resources that can be used in the development of a sustainable building materials industry which will, in return, enhance the welfare of the nation. A map of Botswana is placed at the appendix as reference to various locations in Botswana referred to in this paper.

G Malumbela is with the Botswana International University of Science and Technology, P/Bag 16, Palapye, Botswana (phone: 0027-493-1653; fax: 0027-490-0102; e-mail: malumbelag@biust.ac.bw).

E Masuku is with the Botswana Institute for Technology Research and Innovation, P/Bag 0082 Gaborone, Botswana, (phone: 0027-391-4161; fax: 0027-397-4677; e-mail: emasuku@bitri.co.bw).

II. Cement Industry

The overall requirement of cement by the Botswana construction industry in 2015 was 0.89 million tonnes of which 0.58 million tonnes was imported as ready-made cement and the remaining 0.31 million tonnes was blended by two companies in Botswana [2], [7], [8]. Unfortunately, one of the companies which blended 0.28 million tonnes, imports all its raw materials readily processed. The only processing it does in Botswana is blending, packaging and distribution. The second company, which blended 0.03 million tonnes, imports clinker and gypsum but at least uses limestone and fly-ash from Botswana. The proportion of local raw materials used by this company constitutes only 33% of material used which translates to about 1.1% of the total cement consumed by the construction industry [8]. The approach used by this company is much more beneficial to the nation. Their production rate, however, is very low to make a significant impact in the construction industry. Significantly, the most expensive process in cement manufacturing which is the production of clinker is done outside the country.

Table I shows the cost of 50 kg bag of, arguably, most commonly used cement that is blended locally. As shown in the table, the cost of cement in some areas is near double that in Gaborone. This variation is primarily due to high transportation costs. Note also that the south-east to the eastern region of the country holds major entry points for imported building materials. This price variation therefore cuts across various other imported cements further buttressing the need to set-up local cement production plants.

A desk study of resources for limestone and marble in Botswana which are the primary raw materials in the production of cement showed that potential reserves for these minerals are heavily contaminated with Magnesium Carbonate or Magnesium Oxide [9]-[17]. The few that can be used in the cement industry have insufficient reserves to support the construction industry. For example, the largest reserves for calcitic limestone and marble are Matsiloje limestone at 0.67 million tonnes [14], Mmadinare marbles at 0.68 million tonnes [17] and Makoro marbles at 0.27 million tonnes [16]. These resources are not only widely spaced, and their combined reserves cannot support the industry. Limestone in Matsiloje is currently used as an extender in the production of blended cement [8].

Availability of other raw materials used in cement production such as slag, fly-ash, clay and gypsum was also looked into. Fly-ash is produced at Morupule Power Station and at Sowa Plant. Morupule Power Station produces 0.75 million tonnes of fly-ash annually whilst Sowa Plant produces 0.04 million tonnes of fly-ash annually [18], [19]. Regrettably, less than 0.6% of Morupule fly-ash is currently used in cement production with the rest of the fly-ash dumped in fly-ash ponds. This process is not only overly expensive and it is also a potential hazard to the environment primarily through contamination of underground water resources.

TABLE I
COST OF A TYPICAL 50 KG CEMENT BAG IN BOTSWANA

Location	Cost of 50 kg cement bag (BWP)
Gaborone	49.95
Lobatse	52.75
Francistown	55.96
Tsabong	64.95
Gantsi	69.9
Hukuntsi	74.95
Kasane	79.3
Maun	80
Shakawe	87.5
Charles Hill	95

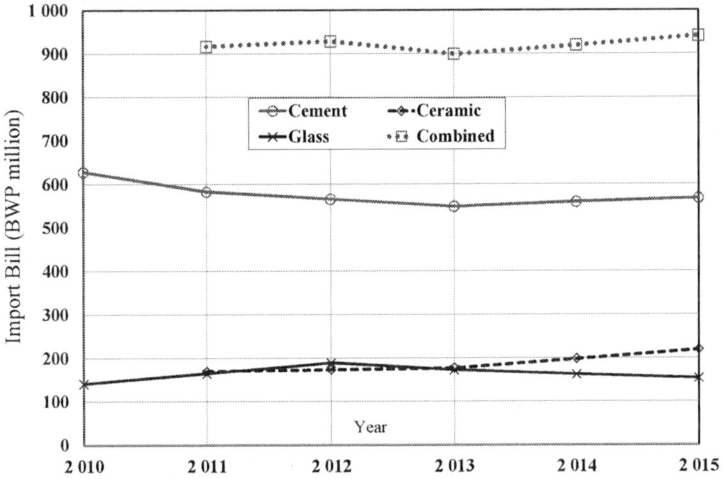

Fig. 1 Import bill for construction materials in Botswana

Slag is produced by Bamangwato Concession Limited (BCL) at a rate of one million tonnes per year [20]. The current stockpile for the slag at BCL mine is in excess of 20 million tonnes. Similar to fly-ash, this slag is not consumed by the industry.

The chemical composition of fly-ash from Morupule B, fly-ash from Sowa and that of slag from BCL is shown in Table II [21]. These materials meet the minimum requirements for use in concrete as stipulated in ASTM C618 and ASTM C989 standards [22], [23]. Fly-ash from Morupule B with a sum of Silicon Dioxide, Aluminum Oxide and Iron Oxide of 53.61% is Class C fly-ash and therefore in addition to having pozzolanic properties, it also has a potential for self-cementitiousness. Fly-ash from Sowa with a sum of Silicon Dioxide, Aluminum Oxide and Iron Oxide of 76.59% is Class F and therefore only has pozzolanic properties.

Large reserves of high quality gypsum are found in Bojanamane at 1.8 million tonnes [24] and in Lebung at 2.4 million tonnes [25]. It can be shown that these reserves are able to support the cement industry for over 150 years. Clay, which is another resource needed in the production of clinker is abundant in many areas of the country. Further discussion on the availability of clay is under the ceramics section.

TABLE II
CHEMICAL COMPOSITION OF FLY-ASH AND SLAG IN BOTSWANA

Raw material	Elements (% by mass)					
	SiO_2	Al_2O_3	Fe_2O_3	CaO	MgO	SO_3
Morupule Ash	29.06	18.89	5.66	17.89	1.58	3.39
Sowa Ash	40.70	29.76	6.13	6.66	1.77	0.98
BCL Slag	34.44	6.55	57.56	2.14	1.70	1.66

Whilst reports on industrial minerals indicate that Botswana does not have sufficient quality carbonates to run a cement manufacturing industry, it undoubtedly has abundant other raw materials needed in the production of cement which are currently not being used. For sustainability of the industry as well as to create employment, it is recommended that import of raw materials be limited only to deficient carbonates but in an unprocessed form. All other processes for cement manufacturing from production of clinker to blending using local resources and packaging cement should be done in the country. Fortunately, Botswana has one of the world largest reserves for coal to supply fuel for the production of clinker [26]. This approach will greatly reduce the import bill and current cost of disposing fly-ash and slag. Most importantly, it will create employment and curb adverse environmental impacts likely to result from these industrial waste.

III. Glass Manufacturing Industry

As already mentioned, the annual import bill for glass and glassware in Botswana is above BWP 140 million [3]. Unlike in the cement industry where primary resources are limited, Botswana has vast quality raw materials for the glass manufacturing industry as summarised in Table III [10]-[13]. Additionally, Sowa Mine produces 0.2 million tonnes of soda ash annually and has the capacity to increase this production to 0.3 million tonnes [19]. Currently this soda ash is exported to South Africa for use in manufacturing glass. According to [27] the rate of returns from the export of soda ash are very low due to high transport costs coupled by that soda ash is a low value commodity. For sustainability of the industry, it is essential that glass manufacturing be done in Botswana. As will be indicated later under Government initiatives, an attempt towards developing a glass manufacturing industry was once made.

TABLE III
Silica Reserves in Botswana

Source	% SiO_2	Reserves
Makuta Hills	99.2	Fairly large reserves
Jakalasi No 1	98,2	Over a million tonnes
Mmajojo Hill	98.4	Over a million tonnes
Shale Hill	97.6	Considerable reserves
Tswapong Hills	97.8	Enormous resources
Suselela Hill	98.64	Fairly large resources
Mmasegomana Hills	98.28	Unlimited resources
Maape village	96.89	Fairly large reserves
Ditsotswane Hill	98.65	Over a million tonnes
Marapoatshwene Hill	97.8	Enormous reserves
Lobatse	97.9	Over a million tonnes
Mmamabula	96.3	Over a million tonnes

IV. Ceramic Industry

Botswana has sizable ceramic raw materials spread countrywide which can be used to produce high quality ceramic products [10]-[13]. Table IV shows a summary of sources of clay with sufficient resources for use at industrial scale. It is worth noting that only three of the possible sources of ceramic raw materials shown in Table IV are known to be used by the ceramic industry to produce clay bricks and, occasionally, pottery. Note that silica sand presented in Table III is also a major raw material for the ceramic industry. The import bill for ceramic products against the available resources clearly suggest that development of a sustainable ceramic industry in Botswana is feasible. This industry would go a long way in job creation in the form of formal employment and engagement in the small, medium and micro enterprises.

TABLE IV
Sources of Clay in Botswana

Sources of clay	Composition	Quantity
Mokgopeetsane River	48.2% sand, 16.6% silt and 35.2% clay	Fairly large reserves
Pilane	53.2% sand, 20.5% silt and 26.3% clay	Considerable reserves
Moshaneng	43 to 54% sand, 21% silt and 24-34% clay	Fairly large reserves
Kgwakgwe Hill Kanye	29% sand, 43% silt, 28% clay	Fairly large reserves
Taung River	34.9% sand, 23.9% silt, 41.2% clay	Fairly large reserves
Gaborone Dam	48% sand, 25.2% silt, 26.8% clay	Fairly large reserves
Metsemotlhaba River	12.7% sand, 32.3% silt, 55% clay	Fairly large reserves
Mokatse	56.7% sand, 12.45 silt, 30.9% clay	Fairly large reserves
Lobatse	44.8% sand, 29.8% silt, 25.4% clay	Considerable reserves
Lekobolo Pan	31.3% sand, 18.1% silt, 50.5% clay	Enormous reserves
Makoro	9.3% sand, 55.7% silt, 35% clay	Considerable reserves
Bonwapitse River	21.4% sand, 20.4% silt and 58.2% caly	Considerable reserves
Shoshong River	14.8% sand, 22.7% silt, 62.5% clay	Considerable reserves
Serule siding	17.8% sand, 31.6% silt, and 50.6% clay	Considerable reserves
Woodhall	16.2% sand, 53.8% silt, and 30% clay	Considerable reserves
Dukwe	1.8% sand, 46.7% silt, and 51.5% clay	Considerable

V. Masonry Units

In Botswana almost all buildings are built from masonry works. These can either be clay or concrete masonry units. Quality of these units is certified by the Botswana Bureau of Standards (BOBS) under the BOS 27 for concrete masonry units and BOS 28 for clay units [28], [29]. Table V gives all masonry products that are currently certified by BOBS [30]. It is unexpected that only five manufacturers have masonry units certified by BOBS whilst each village or town in the country has multiple manufacturers of masonry units. Additionally, the Department of Industrial Affairs has 17 companies registered with them that actively supply masonry units [31], [32]. Note that the Department of Industrial Affairs registers manufacturers with an annual turnover of above BWP 10 million. This therefore indicates that there are many units that are uncertified and yet find use in construction. The safety, reliability and durability of structures built from these are questionable. Work by [32] has pointed to a number of structures in Botswana where premature repairs had to be made and attributed them to masonry units that were not fit-for-purpose.

It is important to note that not only are licensed units scarce, they are also limited to a few locations in the south-east to the eastern part of the country. This implies that for major construction projects in the rest of areas around the country, certified units have to be hauled to those areas. This obviously is at a high cost especially for the furthest areas which can be up to 1800 km away. Table VI shows typical country-wide transport cost of a commonly used clay unit that is certified by BOBS. The nominal cost of this unit without transport is One Pula Thirty Four Thebe (BWP 1.34). It follows then that for some regions such as Shakawe in the north western part of the country, transport cost of this unit can be about 1.6 times more than its nominal cost. This process is certainly not affordable or sustainable.

TABLE V
CERTIFIED MASONRY UNITS IN BOTSWANA

Manufacturer	Location	Scope of Certification
Brickbuild (Pty) Ltd.,t/a Panda (Francistown)	Farm NQ 26, Monarch, Francistown, Botswana	7 MPa concrete masonry units 14 MPa concrete masonry units. 10 MPa concrete masonry units.
Brick Build (Pty) Ltd. t/a Panda (Gaborone)	Plot 20605, Block 3, Broadhurst Industrial, Gaborone, Botswana	10 MPa concrete masonry units. 14 MPa concrete masonry units.
	Dipotsana, Village	10 MPa – Non Face Plastered (NFP) burnt clay masonry units.
Kwena Concrete Products (Pty) Ltd.	Plot 14404/5 Maakgadigau Road, Gaborone West Industrial, Gaborone, Botswana	10 MPa concrete stock bricks 14 MPa concrete stock bricks
Makoro Brick & Tile (Pty) Ltd.	Makoro Village, Botswana	Burnt clay masonry units of classes: Face brick extra (FBX), Face brick standard (FBS) Burnt clay masonry units class: Non Face Brick Extra (NFX)
Lobatse Clay Works (Pty) Ltd.	Plot 4821, Woodhall Industrial, Lobatse, Botswana	Burnt clay masonry units of class: Face bricks extra (FBX) Burnt clay masonry units of class: Face bricks standard (FBS) Burnt clay masonry units of class: NFX

TABLE VI
UNIT COST OF TRANSPORTING TYPICAL LICENSED MASONRY UNIT IN BOTSWANA

Location	Transport cost of 1000 units (BWP)
Lobatse	116
Gaborone	199
Tsabong	803
Hukuntsi	924.99
Francistown	982.55
Gantsi	1202
Charles Hill	1387
Maun	1805
Kasane	1836
Shakawe	2111

The scarcity of certified units in Botswana can be attributed to distribution of developments in the country, availability of raw materials for brick production and Botswana strategies for quality control. The Greater Gaborone area and the Greater Francistown area combined holds over 50% of the Botswana population [33]. These, understandably, are areas where the largest construction takes place and therefore have a higher demand for certified products. It is therefore understandable that they have a greater percentage of certified units. Other probable reasons for the scarcity of certified products are discussed in later sections of this paper.

VI. CONCRETE AGGREGATES

Aggregates needed for concrete and masonry works are mainly crushed rock and river sand. Table VII shows major operational quarries for crushing rocks. Similar to certified units, the majority of quarries are in the south-east to the eastern part of the country. Absence of crushed aggregates in other areas of the country, particularly the Kgalagadi Region, is primarily because they don't have rocks. These areas are mainly covered with vast Kgalagadi sand. Unfortunately, those in areas covered with Kgalagadi sand have to haul crushed aggregates at a high cost. For example, a cubic meter of crushed aggregates costs BWP 275 in Gaborone compared to BWP 975 in Tsabong.

TABLE VII
QUARRIES FOR CONCRETE AGGREGATES IN BOTSWANA

Name	Location	Aggregate
Belabela Quarry	Mma Mashia/ Belabela Farms	Granite
Conduit Investment (Pty) Ltd.	Moshaneng	Crushed Stone
Tamarron Trading One Hundred and Seven	Lobatse	Crushed stone
Quarries of Botswana Radisele Quarry	Radisele	Stone
Quarries of Botswana S/Phikwe Quarry	Selibe Phikwe	Stone
Mmokolodi Quarry	Mmokolodi	Granite
Pioneer Quarries	Moshaneng	Aggregates
Quarries of Botswana Lose Quarry	Lose	Stone
PM Pitlo & Sons	Kuke	Quartz/Feldspar
Nambo Projects	Ghanzi	Stone
Maun Quarries & Brickfield	Maun, Toteng	Crushed Stone
PPC/Kgale Quarries	Gaborone	Stone
Quarries of Botswana. Francistown Quarry	Francistown	Concrete & Road Stone
SA Shippers	Serowe	Aggregates

River sand which is another important raw material for concrete works has been depleted in the Greater Gaborone area to an extent that in 2013 the Botswana Cabinet ordered for the ban of river sand mining in the area. Construction in this area since then has shifted to manufactured sand from crushed rocks. River sand is however still abundant in the central to the eastern region of the country.

One major aggregate resource in Botswana is Kgalagadi sand which covers more than 75% of the country. Unfortunately, unlike river sand and manufactured sand, Kgalagadi sand is very fine with little silt and coarse materials as shown in Fig. 2. The characteristics of Kgalagadi sand make it exhibit behaviour of low cohesion and poor compactibility. These are both undesirable properties of a material for concrete works and for brick moulding. It is not surprising that amongst the list of certified brick products in the country presented in Table V, there are no masonry product that uses Kgalagadi sand. This further explains the scarcity of certified units in some areas of Botswana. In the absence of affordable conventional materials, communities in these area that are underprivileged use the Kgalagadi sand as is in their day-to-day construction works and bear the consequences that are associated with the use of a poor construction material such as excessive cracking and in some instances complete structural failure. Preliminary tests conducted at the Botswana Institute for Technology Research and Innovation (BITRI) indicated that some masonry units made from Kgalagadi sand that find use in the industry have compressive strengths as low as 0.8 MPa [34].

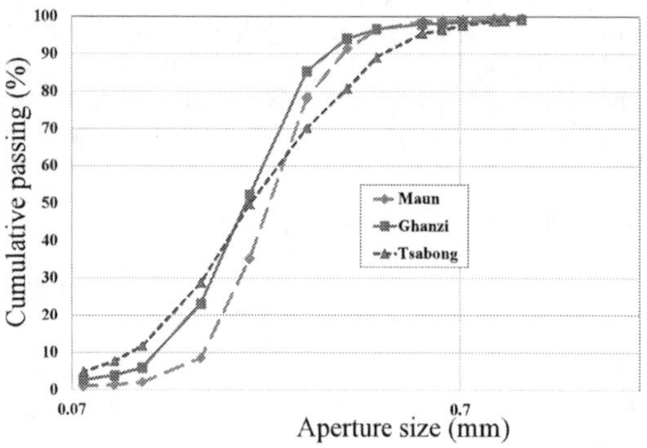

Fig. 2 Sieve analysis of typical Kgalagadi sand

Kgalagadi sand is not only being used by locals in the manufacture of bricks. Almost all construction projects irrespective of the magnitude, use it in concrete works and for mortar. Little research has however been carried out to confirm the quality of works from its use given its unconventional characteristics. The safety and reliability of these structures is therefore questionable.

The review up until now has shown that Botswana has mixed natural resources for the construction industry. Most are in abundance and of required quality for use in the industry and yet are not being used, whilst others are a challenge to use in their natural form. The next section of the present paper looks at governmental initiatives that attempt to address the issue of sustainable building materials in Botswana.

VII. Government Initiatives

The following is a summary of government initiatives that have been put forth to ensure sustainability of the construction industry.

A. Establishment of Botswana Geoscience Institute [26]

This institute was established in 2014 to undertake research in the field of geosciences and to provide specialised geoscientific services. It replaced the Department of Geological Surveys which is mostly cited in this report.

B. Establishment of Botswana Bureau of Standards [30]

BOBS was formed in 1997 with the primary objectives of formulating Botswana standards and co-ordinating quality assurance activities in Botswana with mission to improve the quality of life of the citizens of Botswana. By offering technical services in the areas of standardization, testing of goods and certification of products BOBS play an important role in the construction industry. Unfortunately, the only mandatory product/standard that they have related to the construction industry is cement. All other products are not mandatory leaving it to the manufacturer to certify their product or to a contractor in order to use a certified or non-certified product. To minimise the supply of low strength products, licensing of manufacturers to produce and supply units should be provided only to those manufacturers that meet the BOBS requirements.

C. Establishment of Research Institutions [35]-[37]

The three major research institutions in Botswana that carry out research related to the construction industry are Botswana International University of Science and Technology (BIUST), Botswana Institute for Technology Research and Innovation (BITRI) and University of Botswana (UB). BITRI has developed a technology that uses Kgalagadi sand with a cement blend made from Morupule B fly-ash and Ordinary Portland Cement to produce quality masonry units. Standards for this technology have been developed by BOBS [38], [39]. Currently, BITRI, in collaboration with the Government, is rolling out the technology to areas mentioned in the paper where there are no conventional building materials besides the Kgalagadi sand. Research work by BITRI in collaboration with Botswana Geoscience Institute on manufactured sand and Kgalagadi sand for use in mortar is ongoing.

D. Establishment of Botswana Development Corporation [40]

Botswana Development Corporation (BDC) was founded by the Botswana Government in 1970 as a development finance Institution to promote and facilitate the development of industrial, commercial and agricultural enterprises. Kwena Concrete Products (Pty) Ltd. and Lobatse Clay Works (Pty) Ltd. listed in Table V amongst the four companies that manufacture BOBS certified masonry units were founded through BDC.

In 2008, BDC formed a joint venture with a glass manufacturing company to set-up a 450-tonne float glass manufacturing plant in Palapye. Unfortunately, the project failed during construction stage to meet its target and consequently it was liquidated in 2013. The project has been suspended since then.

VIII. Conclusions

The following is a summary of findings from this review;

1. The import bill for conventional building materials in Botswana is very high and unsustainable.
2. Botswana does not have adequate quality limestone to support a cement manufacturing industry. However, it has abundant other quality cement raw materials that are currently barely used. It is recommended that only limestone be imported but in its raw form with all other processes carried out locally.

3. Botswana has abundant raw materials for ceramic and glass manufacturing industries. It is recommended that these industries be developed to reduce the import bill for building materials and create employment.
4. Despite having adequate quality resources for the production of masonry units, only a limited number of manufactures supply BOBS certified units. It is recommended that BOBS should work closely with the Department of Industrial Affairs to ensure that large suppliers of units mould certified units.
5. Research institutions set-up by Government make a great contribution towards development of a sustainable construction industry in Botswana. The uptake of their work by the construction industry is however very slow.

Appendix

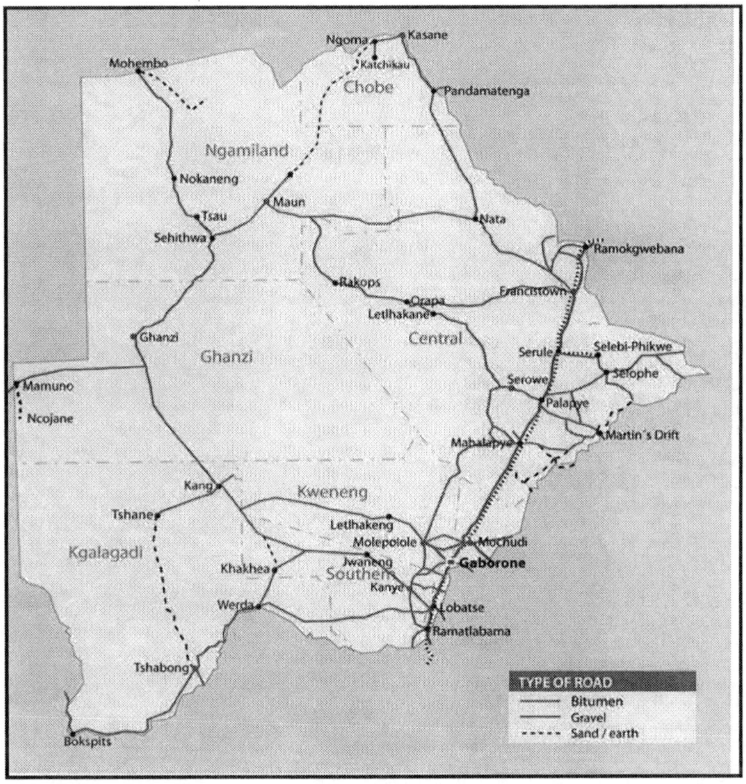

Fig. 3 Map of Botswana

Acknowledgment

The authors thank Botswana International University of Science and Technology and the Institute for Technology Research and Innovation for financial assistance.

References

[1] Statistics Botswana, "Gross Domestic Product," Statistics Botswana, 2016.
[2] Statistics Botswana, "The Imports of Cement from 2010 to 2015 in Pula Value," Statistics Botswana, 2016.
[3] Statistics Botswana, "The Imports of Glass, Glassware, Plastic and Plastic Products from 2010 to 2015 in Pula Value," Statistics Botswana, 2016.
[4] Statistics Botswana, "Annual Importation of Ceramics Tiles from 2010 to 2015 in Pula Value," Statistics Botswana, 2016.
[5] Statistics Botswana, "Botswana Core Welfare Indicators Survey 2009/2010," Statistics Botswana, 2013.
[6] Statistics Botswana, see http://www.cso.gov.bw/ (accessed 30/08/2016).
[7] Pretoria Portland Cement Botswana, private communication, August 2016.
[8] Mastiloje Portland Cement Botswana, private communication, August 2016.
[9] Geological Survey Department, "The Carbonate Resources of Botswana," Department of Geological Survey, 1983.
[10] Geological Survey Department, "A Review of Industrial Minerals in the Gaborone-Lobatse-Kanye – Molepolole Area," Department of Geological Survey, 1985.
[11] Geological Survey Department, "A Review of Industrial Minerals in the Serowe-Palapye-Mahalapye Area Central District," Department of Geological Survey, 1987.
[12] Geological Survey Department, "A Review of Industrial Minerals in North Eastern Botswana," Department of Geological Survey, 1989.
[13] Geological Survey Department, "A Review of Industrial Minerals in the Selibi- Bobonong, Tsetsebjwe, Moeng and Sherwood Areas," Department of Geological Survey, 1996.
[14] Geological Survey Department, "The Final Report on Matsiloje Limestones," Department of Geological Survey, 1990.
[15] Geological Survey Department, "Report on Mahibitswane Carbonate Occurrence," Department of Geological Survey, 1980.
[16] Geological Survey Department, "Report on the Makoro Carbonate

Deposit Central District," Department of Geological Survey, 1981.
[17] Geological Survey Department, "The Establishment of a Cement Industry in Botswana Using Small-Scale Cement Plant Technology," Department of Geological Survey, 1989.
[18] Botswana Power Station, "Sale and Collection of Fly-ash from Morupule 'A' Power Station and Morupule 'B'Power Station," Botswana Power Station, 2016.
[19] Botswana Ash, "Botash Operations (Unpublished data)", Botswana Ash, 2014.
[20] Bamangwato Concession Limited, "Bamangwato Concession Limited (unpublished data)", Bamangwato Concession Limited, 2016
[21] Botswana Institute for Technology Research and Innovation, "Test Report on Chemical Composition of Various Samples of Cement Extenders (unpublished data)," Botswana Institute for Technology Research and Innovation, 2014.
[22] ASTM C 618-12, "Standard Specification for Coal Fly Ash and Raw or Calcined Natural Pozzolan for Use in Concrete," ASTM International, West Conshohocken, PA, USA, 2012.
[23] ASTM C 989-14, "Standard Specification for Slag Cement for Use in Concrete and Mortars," ASTM International, West Conshohocken, PA, USA, 2014.
[24] Geological Survey Department, "Gypsum Reserves at Bojanamane Central District," Department of Geological Survey, 1983.
[25] Geological Survey Department, "Reserve Calculation for Lebung and Bojanamane Gypsum Deposits," Department of Geological Survey, 1984.
[26] Ministry of Minerals Energy and Water Resources, see www.mmewr.gov.bw (accessed 30/08/2016).
[27] Botswana Ash, see http://www.botash.bw/ (accessed 30/08/2016).
[28] Botswana Bureau of Standards BOS 28 -2000, "Burnt clay masonry units – specification," BOBS, Gaborone, Botswana, 2000.
[29] Botswana Bureau of Standards BOS 27 -2000, "Concrete Masonry Units – Specification," BOBS, Gaborone, Botswana, 2000.
[30] Botswana Bureau of Standards, see http://www.bobstandards.bw/ (accessed 30/08/2016).
[31] Department of Industrial Affairs, "Licenced Manufacturing Companies (unpublished)," Department of Industrial Affairs, 2014.
[32] C Nelson, G Malumbela, "On the Compressive Strength of Botswana Masonry Units," Proceedings of the Institution of Civil Engineers Construction Materials, Vol. 167, Issue 5, pp. 271-276, Oct. 2014.
[33] Statistics Botswana, "2011 Botswana Population and Housing Census Alphabetical Index of Districts," Statistics Botswana, 2011.
[34] Botswana Institute for Technology Research and Innovation, "Test of Kgalagadi sand bricks used in the industry (unpublished data)," Botswana Institute for Technology Research and Innovation, 2014.
[35] Botswana Institute for Technology Research and Innovation, see http://www.bitri.co.bw/ (accessed 30/08/2016).
[36] University of Botswana, see http://www.ub.bw/ (accessed 30/08/2016)
[37] Botswana International University of Science and Technology, see http://www.biust.ac.bw/ (accessed 30/08/2016).
[38] Botswana Bureau of Standards BOS 529-1:2013, "Compressed Earth Building Units – Part 1: Kgalagadi Sand Building Units – specification," BOBS, Gaborone, Botswana, 2013.
[39] Botswana Bureau of Standards BOS 529-2:2013, "Compressed Earth Building Units – Part 2: Kgalagadi Sand Building Units – Code of Manufacture," BOBS, Gaborone, Botswana, 2013.
[40] Botswana Development Corporation, see http://www.bdc.bw/ (accessed 30/08/2016).

Experimental Study on Strength and Durability Properties of Bio-Self-Cured Fly Ash Based Concrete under Aggressive Environments

R. Malathy

***Abstract*—**High performance concrete is not only characterized by its high strength, workability, and durability but also by its smartness in performance without human care since the first day. If the concrete can cure on its own without external curing without compromising its strength and durability, then it is said to be high performance self-curing concrete. In this paper, an attempt is made on the performance study of internally cured concrete using biomaterials, namely *Spinacea pleracea* and *Calatropis gigantea* as self-curing agents, and it is compared with the performance of concrete with existing self-cure chemical, namely polyethylene glycol. The present paper focuses on workability, strength, and durability study on M20, M30, and M40 grade concretes replacing 30% of fly ash for cement. The optimum dosage of *Spinacea pleracea, Calatropis gigantea*, and polyethylene glycol was taken as 0.6%, 0.24%, and 0.3% by weight of cement from the earlier research studies. From the slump tests performed, it was found that there is a minimum variation between conventional concrete and self-cured concrete. The strength activity index is determined by keeping compressive strength of conventionally cured concrete for 28 days as unity and observed that, for self-cured concrete, it is more than 1 after 28 days and more than 1.15 after 56 days because of secondary reaction of fly ash. The performance study of concretes in aggressive environment like acid attack, sea water attack, and chloride attack was made, and the results are positive and encouraging in bio-self-cured concretes which are ecofriendly, cost effective, and high performance materials.

***Keywords*—**Biomaterials, *Calatropis gigantea*, polyethylene glycol, *Spinacea oleracea*, self-curing concrete.

I. Introduction

HIGH performance concrete, falling into the category of the so-called modern concrete, is essentially characterized by a cement matrix with low water/cement ratio (w/c), often including mineral additions like silica fume and the use of admixtures as super plasticizers [1]. One of the major problems with these mixtures is their tendency to undergo early-age cracking. While this cracking may or may not compromise the compressive strength of these concretes, it likely does compromise their long-term durability. The phenomenon of early-age cracking is vulnerable and depends on thermal effects, autogenous strains and stresses, drying stress relaxation. Structural detailing and execution was dealt at length [2], [3]. Curing is the critical period of fresh cement in the hardening process, in which concrete develops its fundamental characteristics under specific conditions [4]. A proper curing essentially involves the permanent availability of internal moisture to sustain the hydration reactions at moderate temperatures, and in the absence of external forces at early ages. High performance concrete is particularly sensible to curing problems and may undergo considerable early-age deformations and microcracking development throughout the system [4], [5]. The dense microstructure of modern concrete rapidly develops capillary discontinuity in the still younger porous network. Thus, the access of external water turns out to be unviable to assure continuous saturation of the total bulk volume [6]. Concrete with internal curing may also have evolved from the concept of self-curing concrete, which is based on the introduction of a chemical admixture that is able to reduce water evaporation by a retaining function. The addition of a self-cure chemical based on a water-soluble polymeric glycol leads to improved durability of concrete cured in air [7]. However, the performance of such admixture does not attain the efficiency of the water film curing [8]. The use of water retaining agents should not be seen as internal curing, since it is conceptually based on the internal sealing rather than internal curing. The latter consists of a water-curing agent capable of enhancing cement hydration maintaining optimal curing conditions. Moreover, the selection of a specific curing method is highly dependent on actual design constraints [9]. In either case, there are many examples that require other form of curing procedures rather than the traditional or external curing methods, e.g. high strength concrete applied in submerged pipe systems within oil platform structures. In high performance concrete, capillary discontinuity may significantly delay and even limit the water movement throughout the system. This makes the use of water ponding or other external curing methods inefficient in assuring the continuous internal moisture of the material. In this case, the use of internal curing methods may be required to counteract autogenous deformation. As there is only little experience on the practical use of internal curing methods to improve early-age properties of high performance concrete, the subject constitutes per se a great challenge for technicians and industrialists of the modern age.

The need for internal curing initiates directly from the basic nature of cement hydration process. As a mixture of cement and water reacts to form crystalline and gel hydration products, the water combined into these hydration products generally occupies less space than water in its bulk form. So,

R. Malathy (Professor and Dean R&D) Department of Civil Engineering is with the Sona College of Technology, Salem, Tamilnadu, India (e-mail: malathycivil@sonatech.ac.in).

International Science Index, Structural and Construction Engineering Vol:11, No:3, 2017 waset.org/Publication/10006631

the hydration (and pozzolanic) reactions are accompanied by a net chemical shrinkage as the products occupy less space than the reactants. The hydration of tricalcium silicate reduces about 9.6% by its volume, in the other words, about 0.07 mL/g by considering molar volumes in reaction with 7% on a mass of water needed to mass of C_3S basis [10]. This chemical shrinkage will lead to physical shrinkage of three-dimensional microstructure of cement, before it sets [11], [12]. However, after the cement paste sets and develops a finite resistance to deformation, the chemical shrinkage, in the absence of additional water, may lead to self-desiccation, and pores with partial satisfying may be formed within the microstructure [13]. The pore solution menisci remaining in partially filled pores will create a quantifiable capillary pressure, directly relative to the surface tension of the pore solution and inversely proportional to the size of the largest partially filled pore.

Young's equation as developed by Alberty and Daniels [14] relationship between capillary pressure σ and surface tension of pore solution γ as $\sigma = (-2\gamma\cos\theta)/r$, where θ and r are the contact angle and the pore radius, respectively. From Young's equation, it is clear that, for reducing the capillary pressure, the surface tension needs to be reduced, or the size of the pores has to be increased [1], [2]. In the absence of sacrificial larger pores, the capillary stresses will promptly rise over time, as minor and minor pores within the hydrating cement paste unfilled, whereas the ongoing hydration further reduces the size of the remaining water-filled capillary pores. The reduction in microstructure caused by capillary pressure can be estimated using Mackenzie's modified equation [15]. Early-age cracking caused by the autogenous stresses and strains will create open pathways for ingress of deleterious species. As per the equation given by Kelvin [14], increase in stress occurs due to self – desiccation, and consequently, there will be a reduction in the internal relative humidity of the hydrating cement paste. Hence, to keep the capillary porosity of the hydrating cement paste remaining saturated, a source of available additional water can be provided by the way of internal curing so as to minimize the autogenous stresses and strains. The available additional water will also help to maximize the hydration of cement and mineral admixtures in concrete.

II. Research Significance

The prevention of loss of moisture from concrete is important for the strength development. It is used for the prevention of plastic shrinkage, reduction in permeability, and improvement of resistance to abrasion. The loss in strength after 28 days seems to be directly related to loss of moisture during the first three days. So, internal curing is used as a substitute to overcome such problems. One of the methods to eliminate such problems faced by external curing is the use of lightweight aggregate. The other methods are the use of water absorbent polymers (polyethylene glycol and paraffin wax) and wood derived materials. The disadvantage of using such materials is that, being lightweight aggregate, it can negatively impact strength and can lead to variability in performance. Polyethylene glycols are more controllable but are relatively expensive compared to lightweight aggregate. Wood resultant materials may be substituted with other internal curing materials. It provides consistency at a lower cost, but only 50% strength is reached, and microcracks are introduced by wood derived materials. The above deficiencies can be overcome with bio-materials like *Calatropis gigantea* and *Spinacea oleracia* as curing agents without compromising the strength and durability properties of conventional concrete [16]-[19]. There is significant improvement in strength of fly ash based concretes that are internally cured with these biomaterials [20]. This paper made an attempt on studying the effect of those biomaterials for durability properties of fly ash based high performance concrete.

III. Materials and Mix Proportioning

A. Materials

Ordinary Portland cement OPC 43 Grade confirming to IS: 269-1976 was used throughout the investigation. Class F fly ash from thermal power plant is used in this study. It is light gray in colour, and its specific gravity is 2. Locally available blue granite metal was used for the preparation of concrete. Locally available hard blue granite metal, well graded 20 mm and down size were used. River sand passing through 4.75 mm sieve as per IS383 specifications confirming to zone II were used as fine aggregates. Table I depicts the physical properties of fine and coarse aggregate.

TABLE I
Physical Properties of Fine and Coarse Aggregate

Description	Fine aggregate	Coarse aggregate
Specific gravity	2.6	2.7
Water absorption	1.57%	2.3%
Fineness modulus	3.1(zone II)	6.4
Surface moisture	Nil	Nil
Bulk density	1450 kg/m^3	1750 kg/m^3

B. Self-Curing Agents

The test results concluded that, compared to conventional concrete, concrete with other curing agents, specifically with polyethylene glycol, gave better results. Having arrived at better results with *Spinacia oleracea* and *Calotropis gigantea*, the two curing agents were used in different concrete mix proportions with different quantity of curing agents and tested for compressive strength [16]-[20].

Existing curing agent available in the market is polyethylene glycol, and its molecular weight is between 190 and 210. The specific gravity is 1.12-1.13. Hydroxyl value is 535-590 (mg KOH/g), and pH value is between 5 and 7. *Spinacia oleracea* is usually called as Palak greens in Tamil Nadu and it is a type of green popularly consumed as food product. Curing agent was prepared from the filtrate extract of *Spinacia oleracea* after it was ground well. Its pH value is 6.59. This extract base is added at the time of preparing concrete that is while adding water to the dry ingredients. Placing and compacting the fresh concrete is similar to the conventional concrete, but without curing. The chemical structure shows that it contains (-O-) and (-OH) functional groups. As such, the *Spinacia oleracea* selected as internal curing agent possesses hydroxyl and ether functional group, which is also revealed in Fourier Transform Infra-Red (FTIR) results [20]. *Calotropis gigantea* is a waste plant which grows in fields and terrains without any special care or water. This milk is tried also as another curing agent. Its pH value is found to be 5.17, which is in the range of polyethylene glycol. The method of preparing the extract from the spinach leaves and the collection of milk oozing out of the *Calotropis gigantea* are shown in Figs. 1 (a) and (b).

Fig. 1 (a) Extraction of *Spinaica oleracea*

Fig. 1 (b) Milk from *Calotropis gigantea*

C Mix Proportioning

The mix design for concrete M20, M30, and M40 grades is arrived based on the code IS 10262:2009. The cement is replaced with 30% of fly ash by its weight. The dosage of curing agents was optimized and found to be *Spinacia oleracea* at 0.6% by weight of cement and fly ash, *Calotropis gigantea* at 0.24% by weight of cement and fly ash and polyethylene glycol at 0.3% by weight of cement and fly ash, and strength studies showed encouraging results [16]-[20]. Hence, those proportions were used for making fly ash based concrete samples for durability studies to evaluate long term property of those biomaterials compared with polyethylene glycol.

IV. Experimental Programme

The slump flow tests for workability and compressive strength test as per Indian standards were conducted. The durability properties under aggressive environments like acid attack, alkaline environment, abrasion, and chloride attack were studied.

The acid resistance tests were carried out on cube specimens immersed in water diluted with 1% by weight of sulphuric acid for 45 days continuously. Then, the weight and the compressive strength of the specimens were determined, and the average percentage of loss of weight and the percentage loss of compressive strengths were calculated.

The alkaline resistance tests were carried out on cube specimens which were weighed and immersed in water diluted with 3% sodium hydroxide by weight of water for 90 days continuously. Then, the compressive strength of the specimens was measured, and the average percentages of loss of compressive strengths were calculated.

The Rapid Chloride Ion Penetration test (RCPT) was performed as per ASTM C 1202 to determine the electrical conductance of M 20, M 30, and M 40 grades of mixes at the age of 28 days and to provide a rapid indication of its resistance to the penetration of chloride ions.

V. Results and Discussions

A. Workability

The slump values are shown in Fig. 2, and it is observed that the workability of concrete mixes for all the combinations of the internal curing agents is 4 to 8% lower than the conventionally cured concrete. Even though the slump is less for self-cured concrete when compared to conventionally cured concrete, it is found that concrete with all the curing agents (*Spinacia oleracea*, polyethylene glycol, and *Calotropis gigantea*) satisfies the standard of minimum slump of 50 mm. The slump loss in self cured concrete may be due to the absorption of water, and holding capacity of self-curing agents also may be the reason that total w/b ratio is maintained including the liquid content present in self curing agents.

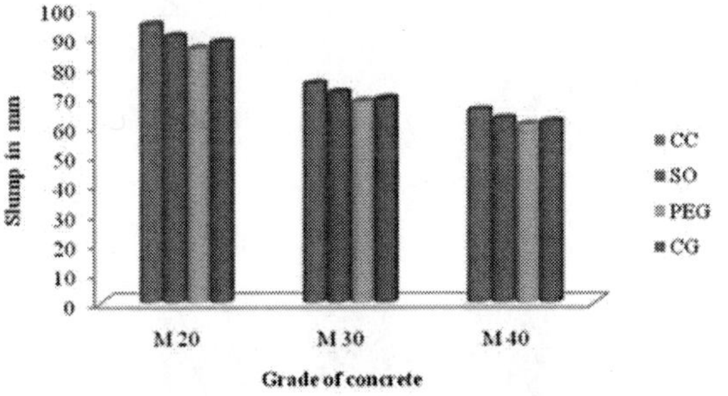

Fig. 2 Slump in mm for various grades of concrete

B. Compressive Strength

The strength activity index is calculated for internally cured concrete keeping unity for compressive strength of conventionally cured concrete at 28 days and it is shown in Fig. 3. The increase in strength at 56 days varies between 8 to 9% higher than that of strength at 28 days for different grades of conventionally cured concrete, whereas the increase in strength found in self-cured concrete is up to 18% because strength contribution by the fly ash is at later ages due to secondary pozzolanic reaction. The presence of ether and hydroxyl functional groups in the concrete enhances the formation of a continuous system of gel which provides better strength development at early ages. Continuous hydration of the mixture at later ages, promoted by the available water due to curing agent, also contributes to the strength increase. Microcracks due to aggregate restraint were minimized since cement paste expanded at early age rather than shrinks and hence improves the strength of concrete.

C. Durability

The results of the acid resistance tests of various concrete mixes at the age of 45 days were observed, and the percentage loss in weight and strength appears in Figs. 4 and 5. Thus, from the test results, it is observed that the attack of acid on concrete is less for concrete cubes with *Spinacia oleracea*. The percentage loss in weight in cubes with *Spinacia oleracea* is around 15% less when compared to conventional concrete and is around 10% and 5% in cubes with polyethylene glycol and *Calotropis gigantea*. Even though the water absorption and porosity values are higher for concrete with self-curing agent, namely *Spinacia oleracea*, the percentage loss in weight and compressive strength due to acid attack is lower. This is due to the fact that the percentage chance for the attack of concrete by acid is less as the pore structure of concrete becomes water tight and does not allow further entry of acid water into the concrete.

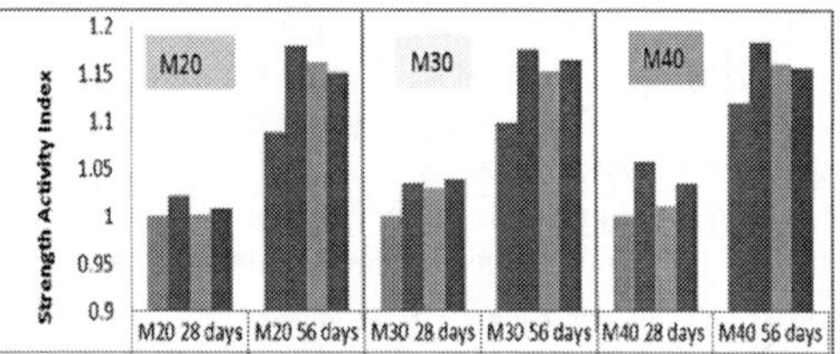

	M20 28 days	M20 56 days	M30 28 days	M30 56 days	M40 28 days	M40 56 days
Conventional concrete	1	1.088	1	1.098	1	1.12
Spinacia oleracea	1.022	1.18	1.036	1.175	1.058	1.184
Calotropis gigantea	1.002	1.162	1.03	1.153	1.011	1.161
Polyethelene Glycol	1.008	1.151	1.04	1.164	1.036	1.156

Fig. 3 Relative compressive strength for various grades of concrete

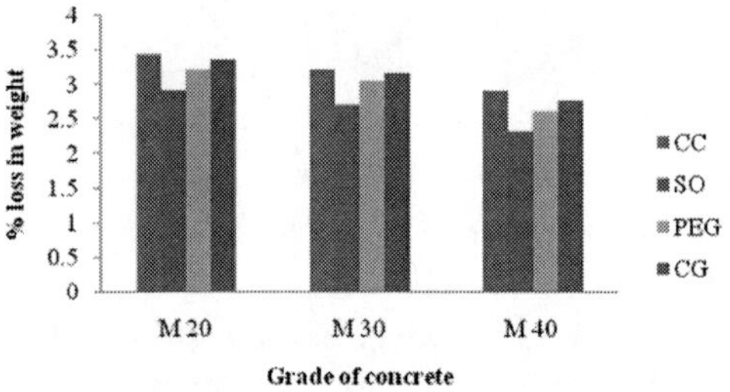

Fig. 4 Percentage loss in weight due to acid attack

Fig. 6 shows the loss of strength due to alkaline attack, and from the figure, it is observed that the concrete with *Spinacia oleracea* curing agent is more resistible against alkaline attack. The loss of strength in cubes with *Spinacia oleracea* is 20% to 30% lower compared to that of conventional concrete. The resistance in *Spinacia oleracea* is higher because further entry of alkaline liquid is prevented, whereas the loss of strength in cubes with polyethylene glycol is around 15% lower compared to conventional concrete. In cubes with *Calotropis gigantea*, the loss of strength is around 30%.

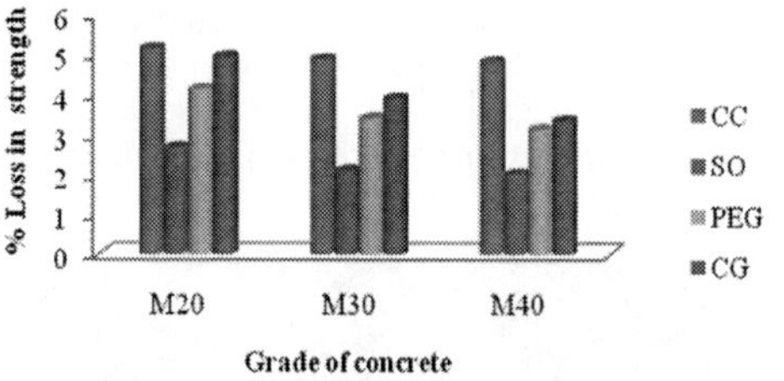

Fig. 5 Percentage loss in strength due to acid attack

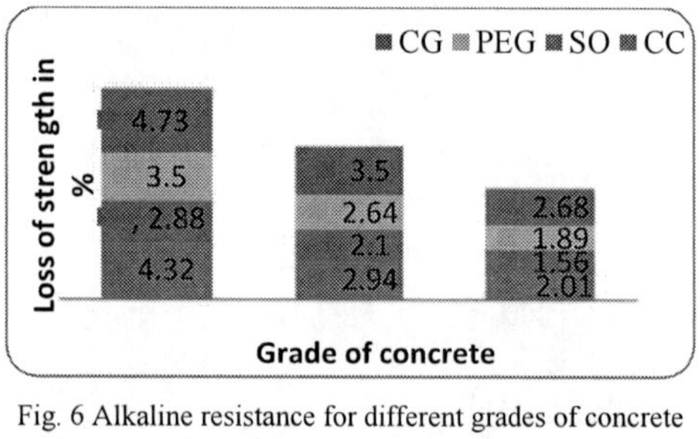

Fig. 6 Alkaline resistance for different grades of concrete

The abrasion resistance test results of various grades and types of concrete at the age of 28 days were observed. Abrasion resistance for different grades of concrete with different curing agents is expressed in percentage wear as shown in Fig. 7.

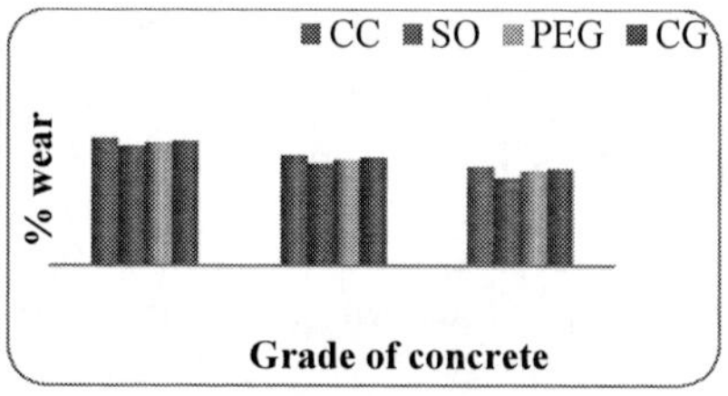

Fig. 7 Abrasion resistance for different grades of concrete

Chloride ion penetration in M 20, M 30, and M 40 grades of concrete for different curing agents is shown in Table II. All the values come under the low and moderate category. When the grade of concrete is higher, there is reduction in the magnitude of charge passed. The reduction in the magnitude of charge passed could be due to the depletion of calcium ions in the gel pore fluids and subsequent reduction of the pH and the development of constricted discontinuous and tortuous pore structure. The quantity of current flowing under the influence of voltage has a direct bearing on the characteristics of the pore solution since the diffusion of ions through the hardened concrete is only due to the electrolytic conduction.

As per ASTM C 1202, the conventional concrete of M 20 and M 30 grade falls under the category of 'Medium' degree of chloride penetrability. The rest of all the other samples falls under the 'Low' degree of chloride penetrability. Fig. 8 shows the specimens, after testing RCPT. Thus, the results of rapid chloride ion penetration test have demonstrated the superior durability characteristics of the mix.

Fig. 8 Specimens subjected to chloride penetration

TABLE II
RAPID CHLORIDE PENETRATION TEST AT THE AGE OF 28 DAYS

Concrete with curing agent	Charge passed as per ASTM equivalent (Coulombs)		
	M20	M30	M40
Conventionally cured concrete	2560	2400	1800
Concrete with *Spinacia oleracea*	1800	1350	1100
Concrete with polyethylene	1600	1440	1050
Concrete with *Calotropis gigantea*	1300	1250	1200

VI. CONCLUSION

From the workability, strength, and durability studies, it is revealed that there is a presence of OH ions in the self-curing concrete. This helps in the effective hydration resulting in better durability properties. It is concluded that the vegetative materials added as internal curing agents perform better workability, strength, and durability characteristics in fly ash based concrete of grades M20, M30 and M40, and such biomaterials as internal curing agents can be used for RCC works, pavements, water tanks, pre-stressed concrete structures without curing with fly ash to achieve long term strength with high performance.

ACKNOWLEDGMENT

Dr. R. Malathy acknowledges University Grants Commission (UGC, New Delhi) for sponsoring this project under Research Award scheme and thanks the Principal and the Management of Sona College of Technology for the encouragement and continuous support towards research.

REFERENCES

[1] Bentz, D.P. Geiker, M.R and Hansen, K.K. "Shrinkage-reducing admixtures and early-age desiccation in cement pastes and mortars", Cement and Concrete Research. Vol.31, No.7, pp.1075-1085, 2001.

[2] Lura, P. Jensen O.M. and Weiss, J. "Cracking in cement paste induced by autogenous shrinkage", Materials and Structures, Vol.42, No.8, pp.1089-1099, 2009.

[3] Shah, S. Weiss, W. and Yang, W. "Shrinkage Cracking - Can It Be Prevented?", ConcreteInternational, Vol. 20, No. 4, pp. 51-55, 1998.

[4] Weber, S. and Reinhardt, H.W. "A New Generation of High Performance Concrete": Concrete with Autogenous Curing, Advanced Cement Based Materials, Vol. 6 pp.59-68, 1997.

[5] Van Breugel, K. "Numerical modelling of volume changes at early ages - Potential, pitfalls and challenges", Material Structure, Vol. 34, No.239, pp.293-301, 2001.

[6] Bentur, A. Igarashi, S. and Kovler, K. "Prevention of autogenous shrinkage in high-strength concrete by internal curing using wet lightweight aggregates", Cement Concrete Research, Vol.31, No.11, pp. 1587-1591, 2001.

[7] Dhir, R.K. Hewlett, P.C. Lota, J.S. and Dyre, T.D. "An investigation into the feasibility of formulating 'self-curing' concrete", Material Structure, Vol.27, No.174, pp.606–615, 1994.

[8] Hewlett, P. "Lea's Chemistry of Cement and Concrete", 4th edition ed: Butterworth-Heinemann, 2003.

[9] Taylor, H.F.W. Cement chemistry, Second edition, Thomas Telford edition published, 1997.

[10] Bentz, D. "Three-Dimensional Computer Simulation of Portland Cement Hydration and Microstructure Development", Journal of the American Ceramic Society, Vol. 80, No.1, pp. 3-21, 1997.

[11] Barcelo, L. Boivin, S. Rigaud, S. Acker, P. Clavaud, B. and Boulay, C. "Linear vs. volumetric autogenous shrinkage measurement: Material behaviour or experimental artefact?", Proceedings of 2nd International Seminar on Self-desiccation and its Importance in Concrete Technology, Lund, Sweden, pp.109-125, 1999.

[12] Sant, G. Dehadrai, M. Lura, P. Bentz, D. Ferraris, C. and Bullard, J. "Detecting the Fluid-to-Solid Transition in Cement Pastes: Part I - Assessment Techniques", Concrete International, Vol.31, No.6, pp.53-58, 2009.

[13] Lura, P. Couch, J. Jensen, O. and Weiss, W. "Early-Age Acoustic Emission Measurements in Hydrating Cement Paste: Evidence for Cavitation during Slidification Due to Self Desiccation", Cement and Concrete Research, accepted for publication, Vol.39, No.10, pp.861-867, 2009.

[14] Alberty, R. and Daniels, F. Physical Chemistry. New York: John Wiley & Sons, p. 682, 1980.

[15] MacKenzie, J.K. "The elastic constants of a solid containing spherical holes", Proceedings of the Physics Society, Vol.683, pp.2-11, 1950.

[16] Geetha, M. and Malathy, R. "Comparison of strength properties of concrete on addition of admixtures", ISSN 0975-6744, Vol.1, No.1, pp. 7-10, 2009 (e-journal).

[17] Geetha, M. and Malathy, R. "Comparison of strength properties of concrete on addition of admixtures", International Journal of computer Applications in Engineering, Technology and Science (IJ-CA-ETS),ISSN 0974-3596/, Vol.2, No.1 , pp.130-133, 2009

[18] Geetha, M. and Malathy, R. (2010), 'Durability properties of self curing concrete by addition of vegetative material as Admixtures' in Technology today , Quarterly journal ISSN 2180-0987, March 2010 ,Volume II issue -1, pp 99-104.

[19] Geetha, M. and Malathy, R. (2011), 'Comparative strength and durability properties of Polymeric materials as Self curing agents', journal of International Science and Technology, ISSN : 0975- 5462, Volume 3 , issue 1,pp 776-771

[20] R. Malathy, "Effect of bio materials as self-curing agents in fly ash based concrete", Proceedings of Advancements in structural concrete , ACECON 2015, Vol.I, pp. 177-184.

R, Malathy graduated in the year 1992, obtained M.E. (Structural Engineering) in 1998 both from Madras University, Chennai and Ph.D in 2005 from Bharathiyar University, Coimbatore. She has more than 25 years of teaching experience and 11 years in research. She is the principal investigator for 6 R&D projects (funded by DST, AICTE, IEI , UGC etc.,) worth more than 60 lakhs. She has done extensive research work on High Performance Concrete, Self Compacting Concrete, Self Curing Concrete, Geo Polymer Concrete, Papercrete and Effect of mineral and chemical admixtures on properties of Concrete. She published about 100 papers in reputed journals and conferences at international and National levels. Her recent researches are utilization of steel slag in concrete for various applications like pavements, precast prestressed elements and RCC elements. For her innovative research work on utilization of steel slag, JSW Steels, Macheri, honoured with an award on 15th August 2016. She is a member of various professional bodies such as IEI, ISTE, ICI, ACI (India chapter) and TNSF. Now she is the Chairman of Indian Concrete Institute ICI – Salem Centre and GC member of ICI. She received Best Teacher Award for the year 2004 in Kongu Engineering College and Innovative Teacher Award for the year 2005 from CLHRD, Mangalore. She got "Career Award for young Teachers" for the year 2006 from AICTE, New

Delhi. She also received "Best Paper Award" in the year 2006 published in ICI journal. She patented a product "Electronic Level Indicator using Sensors" in the year 2006 and now 3 patents are in 2011. She organized several workshops and conferences and delivered lectures on Concrete Technology, Effective Teaching Learning Process, Human Resource Management and Earthquake Engineering in various institutions and to practicing Engineers. She visited Singapore to present a paper on "Efficiency factor for silica fume and metakaloine" in International Conference on Our World in Concrete and also chaired one session. She visited South Korea on 24th September to present a paper on Geo Polymer Concrete in the 12th International Congress on Polymers in Concrete. She also visited London to present papers in International Conference and she has been invited as a resource person in the Conference on Building Materials in Germany. She received Best Engineering College Teacher Award from ISTE (TNC) and Best Women Engineer Award from Institution of Engineers in the year 2008. She has been recognized for inclusion as a biographee in the 2009 edition of Marquis' Who's Who in the World, a globally distributed publication continues to be recognized internationally as the premier biographical data source of notable living individuals from every significant field of endeavor. Under her guidance, Project students received Socio Engineer Award for the project regarding "treatment of dyeing factory effluent by using fly ash as an adsorbent". She received Best Senior Engineer award 2011 from Salem Chapter of IEI. She received two National awards from ISTE one is for her outstanding research work on development of eco friendly, low cost building material and another one is for Best woman engineering college teacher for the year 2012. She has produced 11 doctorates and guiding 7 Ph.D scholars. Recently she has been honoured with prestigious Research Award by UGC, New Delhi (2014-2016). Presented two papers in International Conference at Hong Kong in October 2015. Presently she is the Professor & Dean (Research & Development) in Sona College of Technology (Autonomous) Salem, Tamilnadu, India and also established and heading the research centre SONA COIN (COncrete INnovation).

World Academy of Science, Engineering and Technology
International Journal of Computer, Electrical, Automation, Control and Information Engineering Vol:11, No:1, 2017

Multi-Level Meta-Modeling for Enabling Dynamic Subtyping for Industrial Automation

Zoltan Theisz, Gergely Mezei

***Abstract*—**Modern industrial automation relies on service oriented concepts of Internet of Things (IoT) device modeling in order to provide a flexible and extendable environment for service meta-repository. However, state-of-the-art meta-modeling techniques prefer design-time modeling, which results in a heavy usage of class sometimes unnecessary static subtyping. Although this approach benefits from clear-cut object-oriented design principles, it also seals the model repository for further dynamic extensions. In this paper, a dynamic multi-level modeling approach is introduced that enables dynamic subtyping through a more relaxed partial instantiation mechanism. The approach is demonstrated on a simple sensor network example.

***Keywords*—**Meta-modeling, dynamic subtyping, DMLA, industrial automation, arrowhead.

I. Introduction

THE meta-model based industrial automation frameworks could be the next wave of solutions to meet the challenges set by Industry 4.0. The open IoT infrastructure requires a fully service oriented conceptualization of industrial automation. Although the service layer interface hides the internal data structure of concrete embedded devices, data modeling still remains an important part of the final solution. The services must be registered in a flexible repository and they shall be available for look-ups. Therefore, the necessary modeling efforts must go well beyond the current static nature of state-of-the-art model building. Clearly, object-oriented programming or meta-model facilitated subtyping supports only design time modeling. Although traditional industrial automation solutions could be easily satisfied by a fixed categorization of embedded devices, the newer wave of SOA-based automation frameworks openly embrace System of Systems (SoS) design concepts of composability. Hence, corresponding modeling solutions must be researched and made available. A straight-forward candidate solution could be dynamic subtyping, which provides both the benefits of dynamic introduction of new types and the consistency of already defined types within the (sub-)systems.

In this paper, Section II introduces the Arrowhead framework, then Section III explains dynamic subtyping. Next, in Section IV, a dynamic multi-level meta-modeling approach is introduced in detail. Then, Section V shows how the approach can be applied to come up with a viable implementation for dynamic subtyping, which is also demonstrated through a simple example in Section VI. Finally, Section VII concludes the paper.

Zoltan Theisz is with Evopro Innovation, Budapest, Hungary (e-mail: zoltan.theisz@evopro.hu)

Gergely Mezei is with the Department of Automation and Applied Informatics at the Faculty of Electrical Engineering and Informatics of Budapest University of Technology and Economics, Budapest, Hungary (e-mail: gmezei@aut.bme.hu).

II. The Arrowhead Framework

The Arrowhead framework [1] aims to become one of the best technology answers currently available to the general challenges raised by the Industry 4.0 revolution in industrial automation. The framework is based on the concept of the IoT, it clearly follows the principles of the SoS, and it has been implemented along the well-known guidelines laid out by Service Oriented Architecture (SOA). All these individual facets have been united to facilitate collaborative industrial automation of networked embedded smart devices. The concept of a service within the framework is defined as a piece of information that can be exchanged between or among participating communication systems. Since the final goal of the framework is to provide SoS features, it must also support a great variety of different sensors and activators, or any arbitrarily combinations thereof, through unique and well defined set of registered interfaces. The services are reusable and can be provided via various systems by explicitly declaring them via communication profiles each possessing the following three main characteristics: transfer protocol, security mechanism, and data format. Within the framework, the concept of a system is defined as a composable entity (according to the well-known principles of SoS) that can produce and consume services. The system that is producing a service is labeled as the Service Provider, while the system that is consuming a service is referred to as the Service Consumer. Also, a particular Arrowhead system may play the role of a provider of one or more services or a consumer of some other services, or even both at the same time.

Core Arrowhead mechanisms make use of three mandatory framework-wide services: Service Registry, Orchestrator and Authorization. These three basic services combined with the above definition of components within the system(s) establishes an SoS adapted version of SOA principles in order to enable a new wave of industrial automation facilitated by smart embedded devices. As can be seen in Fig. 1, firstly, the Service Registry stores all meta-data of any registered services and provider systems. Secondly, Management sets up and orchestrates the communication channels in order to combine the services into application systems, eventually also incorporating their reconfigurability. Finally, Authorization ensures that only properly authorized systems may provide and/or consume services within the framework.

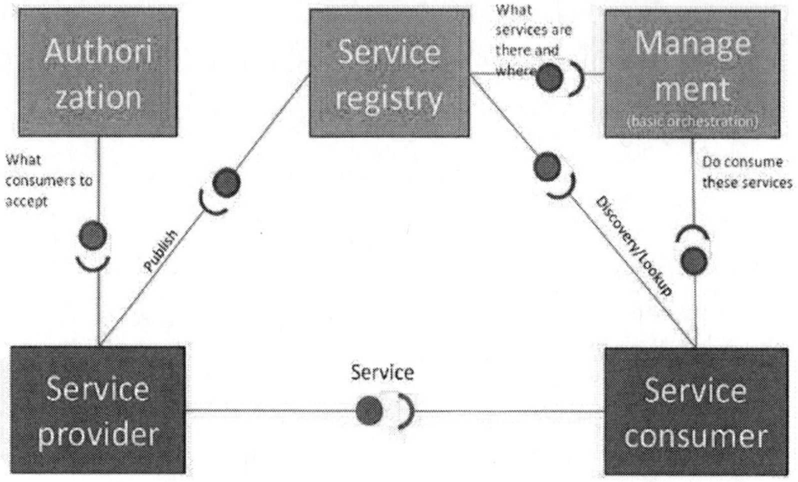

Fig. 1 Arrowhead's interpretation of SoS adapted SOA

III. Dynamic Subtyping

A. Need for Dynamic Subtyping

The Service Registry is the central modeling repository of the Arrowhead framework: it ties together both design-time and run-time aspects of service models covering the full life cycle of any particular service an industrial automation system may consist of. Hence, it is of paramount importance that it keeps on remaining both extendable and flexibly constrained. Thus, only proper services such as new compatible sensor instances are allowed to be inserted in or deleted from. On one hand, proper subtyping shall be needed to constrain the meta-data structure of the registered instances in the repository. On the other hand, it must be practical in the sense that new types may be defined dynamically by step-by-step addition of new or removal of already existing structural items in the defined meta-data structure. Moreover, sometimes particular service meta-data instances may be reused as templates for subtypes that extend those concrete values with further data structures. Hence, although legacy meta-model based subtyping is an effective workhorse for any object-oriented repository implementation, it may become very difficult to provide an extendable and type-correct internal mechanism for its opening towards the needs of Arrowhead compatible SoS based SOA.

B. Multi-Level Instantiation

Classical object-oriented modeling principles effectively restrict the number of meta-modeling levels to two: one of them being the class level where the data structure is defined, and the other one being the object level where those instances are stored which comply to the data structures. This restriction ultimately means that subtyping must be a "vertical" relation between the entities at the class level, while instantiation connects the class level entities to their corresponding instances at the object level. An unfortunate consequence of this strict isolation is that the class level gets frozen in design-time and cannot be further extended by new entities in run-time. However, in a multi-level meta-modeling architecture instantiation are allowed to take place across multiple levels. These techniques usually distinguish between two options: shallow instantiation means that the structural information is defined on the nth modeling level and it is used on the immediate instantiation (n+1)th level, while deep instantiation allows defining the information on the nth modeling level and use it on the (n+x)th ($x > 0$) modeling level [2]. Object-oriented instantiation principles support only shallow instantiation, which results in a rigid and unnecessary separation between class level and object level. This is the reason behind the fact that subtyping and instantiation are split and cannot be directly related to each other, one being "horizontal" and the other "vertical". Hence, legacy subtyping is static and instantiation is dynamic in nature, respectively. Therefore, dynamic subtyping should combine "horizontal" and "vertical" aspects by providing a means to be able to introduce new attributes and operations to the existing model entities in a precisely controlled manner. In a sense, the new dynamic subtyping concept will become a "horizontal" instantiation, in other words, it is carried out through the insertion of an extra meta-level.

IV. Dynamic Multi-Level Algebra

Dynamic Multi-Layer Algebra (DMLA) [3]-[5] is the theoretical framework facilitating the introduction of dynamic subtyping. It consists of three constituent parts: the first one defines the modeling structure and also establishes the ASM [6] functions declared on this structure. The second part creates the initial set of modeling constructs, the so-called minimal set of built-in entities which are needed to bootstrap any further modeling. Finally, the third part validates and also safeguards the correctness of the instantiation mechanism.

A. Modeling Structure

DMLA represents the model as a Labeled Directed Graph (LDG), where each node and edge can have labels attached. Those labels define the attributes of the model elements of the LDG. For simplicity, we apply dual field notation in labeling for Name/Value pairs. The data representation of the model elements is defined by the below labels:

- X_{Name}: name of the model element
- X_{ID}: globally unique ID of the model elementv
- X_{Meta}: ID of the meta-model definition
- $X_{Cardinality}$: cardinality of the model element
- X_{Value}: value of the model element
- $X_{Attributes}$: list of attributes

Now, let us define the algebra over those labels. The superuniverse $|\mathfrak{A}|$ of a state $\mathfrak{A}$ of the Multi-Layer Algebra consists of the following universes:

- U_{Bool}: containing logical values {true/false}
- U_{Number}: containing rational numbers and a special symbol ∞ representing infinity
- U_{String}: containing character sequences of finite length
- U_{ID}: containing all the possible entity IDs
- U_{Basic}: containing elements from { U_{Bool} U U_{Number} U U_{String} U U_{ID} }

Additionally, all universes contain a special element, undef, which refers to an undefined value. The labels of the entities take their values from the following universes:

- X_{Name}: U_{String}
- X_{ID}: U_{ID}
- X_{Meta}: U_{ID}
- $X_{Cardinality}$: [U_{Number} , U_{Number}]
- X_{Value}: U_{Basic}
- $X_{Attributes}$: U_{ID} []

The label Attrib is an indexed list of IDs, which refers to other entities. Now, let us have a simple example:

SensorID=69, SensorMeta=42, SensorCardinality=[0, inf], SensorValue=undef, SensorAttrib=[]

The above definition describes a Sensor entity with its ID set to 69, and the ID of its meta-model being 42. From now on, the same semantics will be shown in a more compact representation (both tuples and lists by square brackets).

{"Sensor", 69, 42, [0, inf], undef,[]}

B. ASM Functions

ASM functions are used to control how one can change the states within the ASM. DMLA relies on both shared and derived functions. Shared functions show the current attribute configuration of a model entity. DMLA allows the values of these functions to be modified either by the algebra or by its environment. Derived functions represent calculations which cannot change the model; they only obtain or restructure existing entity information. The vocabulary $\sum$ of DMLA consists of the following shared functions (for technical details see [5]):

- *Name(ID)*
- *Meta(ID)*
- *Card (ID)*
- *Attrib(ID,Idx)*
- *Value(ID)*
- *Contains(ID_1,ID_2)*
- *DeriveFrom(ID_1,ID_2)*

C. Bootstrap Mechanism

The ASM functions have defined the basic structure of the algebra and the means to query and change the model. However, some initial built-in constructs are also needed for any practical modeling. As a minimum, the bootstrap creates the basic types and the so-called principal entities. The basic types correspond to DMLA's defining universes such as Basic, Bool, Number, String, and ID. Besides these, the first two types of like entities are also defined: Attribute and Base. They act as root entities of all other attributes and node meta-types, respectively. Furthermore, a third entity, AttribType, is also defined that restricts, by corresponding validation formulae, the value of the attributes within the instances. Namely, the label Value of an AttribType specifies which type may be used within the instance given by the referred attribute. However, AttribType and its Value field must only be filled in if the given attribute has been instantiated; otherwise, AttribType can be totally left out. Nevertheless, AttribType can only be applied for attributes.

The principal entities of DMLA are defined as follows:

```
{"Attribute", IDAttribute, IDAttribute,
[0,inf], undef,
[
 {"Attributes", IDAttributes, IDAttribute,
   [0, inf], undef,[]}
]}


{"Base", IDBase, IDBase,
[0, inf], undef,
[
 {"Attributes", IDAttributes, IDAttribute,
   [0, inf], undef,[]}
 {"Links", IDLink, IDAttribute,
   [0, inf], undef,[]}
]}


{"AttribType", IDAttribType,IDAttribute,
[0,1],undef,
[
 {"AType", IDAType, IDAttribType,
   [0,1],IDID,[]}
]}


{"Node", IDNode, IDBase,
[0, 0], undef,
[
 {"Attributes", IDAttributes, IDAttribute,
  [0, inf], undef,[]}
]}
```

D. Validation Formulae

During the instantiation process, only those instances are allowed to be created that do not violate any constraints set by the meta-definitions. Thus, DMLA distinguishes between valid and invalid models, where validity checking is based on formulae. Hence, it is assumed that whenever external actors change the state of the algebra, those formulae must be evaluated. This dynamic property enables DMLA to be applied both in design-time and run-time. The mathematical details of the formulae are published in [4].

Validation formulae take an Instance entity and a MetaType entity, and then check if the Instance entity is a valid instance of the MetaType entity. However, the formula φMeta takes only one parameter and validates if the given entity has enough valid instances according to the cardinality label.

- $\varphi_{IsInstantiated}(C, I)$: checks if *I* is instantiated in *C*
- $\varphi_{InstCounter}(C, a_1, a_2)$: checks if two attributes in the same container *C* are originated from the same meta definition
- $\varphi_{CardinalityCheck}(C, I)$: checks if an attribute *I* violates the cardinality constraints in the container *C*
- $\varphi_{Typecheck}(T, v)$: checks if a specific value *v* violates the type constraint *T*
- $\varphi_{AttribCheck}(I, a)$: checks if an attribute *a* is a valid instantiation of a meta attribute of the meta of container *I*
- $\varphi_{LabelCheck}(I, M)$: checks the correct usage of *Meta* label
- $\varphi_{EntityIns}(I, M)$: checks if at least one of the attributes is instantiated, or has its value set
- $\varphi_{AttribSrc}(I, M)$: checks if there is an attribute violating the cardinality constraint or not having a valid meta definition in the meta of the entity
- $\varphi_{ValueCheck}(I, M)$: checks the validity of the *AttribType* definitions
- $\varphi_{Link}(I, M)$: checks if instances of *Base* have the correct number of links
- $\varphi_{Meta}(M)$: checks the cardinality limits
- $\varphi_{IsValid}(I, M)$: checks the validity of the new ASM state is by combining $\varphi_{LabelCheck}(I, M)$, $\varphi_{AttribSrc}(I, M)$, $\varphi_{EntityIns}(I, M)$, $\varphi_{AttribType}(I, M)$ and $\varphi_{Link}(I, M)$

V. Dynamic Subtyping via DMLA

DMLA provides the means for dynamic subtyping by simply constraining the generic instantiation mechanism to a special subcase. Namely, dynamic subtyping is only allowed to modify the structure of the meta-entity, but no value substitution must take place there. More precisely, the original formula of φAttribCheck(I, a) used to be:

$\varphi_{AttribCheck}(I, a)$:

$\varphi_{IsValid}(a, Meta(a)) \lor$

$\exists j: Attrib(Meta(I), j)=a \lor$

$\{\exists m, k: m=Attrib(Meta(I), k) \land Value(m)=Value(a) \land Name(m) = Name(a) \land Meta(m) = Meta(a) \land (\nexists i: Attrib(a,i) \neq Attrib(m,i))\}$

The formula has three parts which check if an attribute is:

- either a valid instantiation
- or a copy
- or a meta

of a meta attribute of the meta of container *I*. Therefore, if this formula is slightly modified it can easily grasp the essence of dynamic subtyping in the following way:

$\varphi_{AttribCheck'}(I, a)$:=

$\{\exists m, k: m=Attrib(Meta(I), k) \land Value(m)=Value(a) \land Name(m) = Name(a) \land Meta(m) = Meta(a) \land (\nexists i: Attrib(a,i) \neq Attrib(m,i))\}$

In other words, dynamic subtyping only copies existing structures or introduces new structures if related cardinality constraints allow that. Hence, DMLA can be easily incorporate dynamic subtyping by restricting its original instantiation formula.

VI. Example

In order to showcase the subtyping and the instantiation mechanism of DMLA, let us examine a simplified Arrowhead network device scenario. In this scenario, an "abstract" SensorType service entity is defined that allows an arbitrary number of wireless communication channels for data collection and also provides a single IP address for network traffic.

```
{"SensorType", IDSensorType, IDNode,
 [0, inf], undef,
 [
 {"WirelessChannel",IDWirelessChannel,IDAttribute,
   [0,inf],undef,
   [
    {"WCType",IDWCType,IDAttribType,
    [0,inf], IDString,[]}
   ]
  }
 {"IPAddr",IDIPAddr,IDAttribute,
   [1,1],undef,
   [
    {"IPType",IDIPType,IDAttribType,
    [0,inf], IDString,[]}
   ]
  }
 ]}
```

This SensorType service entity is only a meta-type definition; therefore, a service must first instantiate it, for example, by assigning a particular IP address to its network interface, in order for it to be registered into the Service Registry.

```
{"MySensor", IDMySensor, IDSensorType,
 [0, inf], undef,
 [
 {"WirelessChannel",IDWirelessChannel,IDAttribute,
   [0,inf],undef,
   [
    {"WCType",IDWCType,IDAttribType,
    [0,inf], IDString,[]}
   ]
  }
 {"IPAddress", IDIPAddress, IDIPAddr,
  [1,1], "192.168.0.1", []
  }
 ]}
```

In a classical subtyping implementation, the type of the service cannot change any longer, but if dynamic subtyping has been enabled, then

- either a new "concrete" type can be derived from SensorType
- or the MySensor instance can dynamically change its structure

by changing the cardinality of WirelessChannel. Let us assume that entity SpecificSensor must have at least five wireless channels and the maximum amount is eight. In the first case, the result looks like as follows:

```
{"SpecificSensor", IDSpecificSensor, IDSensorType,
 [0, inf], undef,
 [
 {"WirelessChannel",IDWirelessChannel,IDAttribute,
   [5,8],undef,
   [
    {"WCType",IDWCType,IDAttribType,
    [0,inf], IDString,[]}
   ]
  }
 {"IPAddr",IDIPAddr,IDAttribute,
   [1,1],undef,
   [
    {"IPType",IDIPType,IDAttribType,
    [0,inf], IDString,[]}
   ]
  }
 ]}
```

In the latter case, the definition is the following:

```
{"MySensor", IDMySensor, IDSensorType,
 [0, inf], undef,
 [
 {"WirelessChannel",IDWirelessChannel,IDAttribute,
   [5,8],undef,
   [
    {"WCType",IDWCType,IDAttribType,
    [0,inf], IDString,[]}
   ]
  }
  {"IPAddress", IDIPAddress, IDIPAddr,
  [1,1], "192.168.0.1", []
  }
 ]}
```

Note that the ID of the entity remains the same; however, the new structure allows it to be further instantiated, hence the registered service can now spawn sub-services within the set cardinality limits.

VII. Conclusion

Object-oriented modeling artificially limits subtyping to the design time domain only. Although traditional industrial automation solutions used to live with fixed categorization of embedded device types, the new wave of SOA based automation frameworks openly embrace SoS concepts and thus open up the up till now closed world of available sensors and actuators. Hence, a corresponding modeling approach must be put in place in order to facilitate controlled dynamic subtyping that may be lead to higher flexibility without any incurred backdrop on type precision. In this paper, we have proposed a candidate solution which is based on a simple relaxation of a formula of the multi-level meta-modeling approach called DMLA. The applicability of the proposed solution has been showcased by a simple sensor example. The practical implementation of the solution is still work in progress, though it may later become a standard element of any model based implementation of the Service Registry used by the Arrowhead framework.

References

[1] F. Blomstedt, "Arrowhead Framework Cookbook," 2014.

[2] C. Atkinson és T. Kühne, „The Essence of Multilevel Metamodeling," The Unified Modeling Language. Modeling Languages, Concepts, and Tools, Volume 2185, pp. 19-33, 2001

[3] Z. Theisz, G. Mezei, „An Algebraic Instantiation Technique Illustrated by Multilevel Design Patterns," in MULTI@MoDELS, Ottawa, Canada, 2015.

[4] Z. Theisz és G. Mezei, „Multi-level Dynamic Instantiation for Resolving Node-edge Dichotomy," in Proceedings of the 4th International Conference on Model-Driven Engineering and Software Development, Rome, Italy, 2016.

[5] Z. Theisz és G. Mezei, „Towards a novel meta-modeling approach for dynamic multi-level instantiation," in Automation and Applied Computer Science Workshop, Budapest, Hungary, 2015.

[6] E. Boerger and R. Stark, Abstract State Machines: A Method for High-Level System Design and Analysis, Springer-Verlag Berlin and Heidelberg GmbH & Co. KG, 2003.

Electrophoretic Deposition of p-Type Bi_2Te_3 for Thermoelectric Applications

Tahereh Talebi, Reza Ghomashchi, Pejman Talemi, Sima Aminorroaya

***Abstract*—**Electrophoretic deposition (EPD) of p-type Bi_2Te_3 material has been accomplished, and a high quality crack-free thick film has been achieved for thermoelectric (TE) applications. TE generators (TEG) can convert waste heat into electricity, which can potentially solve global warming problems. However, TEG is expensive due to the high cost of materials, as well as the complex and expensive manufacturing process. EPD is a simple and cost-effective method which has been used recently for advanced applications. In EPD, when a DC electric field is applied to the charged powder particles suspended in a suspension, they are attracted and deposited on the substrate with the opposite charge. In this study, it has been shown that it is possible to prepare a TE film using the EPD method and potentially achieve high TE properties at low cost. The relationship between the deposition weight and the EPD-related process parameters, such as applied voltage and time, has been investigated and a linear dependence has been observed, which is in good agreement with the theoretical principles of EPD. A stable EPD suspension of p-type Bi_2Te_3 was prepared in a mixture of acetone-ethanol with triethanolamine as a stabilizer. To achieve a high quality homogenous film on a copper substrate, the optimum voltage and time of the EPD process was investigated. The morphology and microstructures of the green deposited films have been investigated using a scanning electron microscope (SEM). The green Bi_2Te_3 films have shown good adhesion to the substrate. In summary, this study has shown that not only EPD of p-type Bi_2Te_3 material is possible, but its thick film is of high quality for TE applications.

***Keywords*—**Electrical conductivity, electrophoretic deposition, p-type Bi_2Te_3, thermoelectric materials, thick films.

I. Introduction

ALMOST 60% of the energy produced by mankind is wasted in the form of unusable thermal energy in various systems such as power plants, cars, planes, computers, or even stovetops. If we can utilise this enormous amount of energy, we can reduce fossil fuel consumption and greenhouse gases, as well help to decrease the effects of global warming [1], [2]. TE materials are especially attractive in this regard because they can convert waste heat into electricity. Using TE materials can increase the overall efficiency of the system by several percentage points [3], [4]. In addition, TEG are reliable and silent with no moving parts [5]. Although TEGs are currently being used in limited applications such as deep-space satellites, high temperature heat-waste recovery and remote power generation for unmanned systems, there remain a number of challenges which need to be addressed before the commercialization of these devices can occur [6], [7].

T. Talebi is with the Mechanical Engineering School, the University of Adelaide, South Australia, Australia (phone: 883-133-360; fax: 883-134-367; e-mail: tahereh.talebinamakrodbari@adelaide.edu.au).

R. Ghomashchi is with the Mechanical Engineering School, the University of Adelaide, South Australia, Australia (e-mail: reza.ghomashchi@adelaide.edu.au).

P. Talemi is with the Chemical Engineering School, the University of Adelaide, South Australia, Australia (e-mail: pejman.talemi@adelaide.edu.au).

S. Aminorroaya is with Institute for Superconducting & Electronic Materials, University of Wollongong, New South Wales, Australia (e-mail: sima@uow.edu.au).

One of the challenges is the high production cost of TEGs. In addition, the devices are too heavy especially for mobile applications and their specific power to weight factor is low. Therefore, EPD as a cost-effective film fabrication method has been utilized to address these two vital issues for TE applications.

EPD is a film fabrication technique from colloidal suspensions [8]. In EPD, a DC electric field is applied to a suspension of charged particles enabling them to be attracted to the conductive substrate with the opposite charge and form a thick film [9].

EPD is a fast and cost-efficient method, requiring simple equipment. It has the ability to fabricate high quality and crack-free films with high degree of uniformity in the microstructure [10]. It is scalable for industrial applications and its setup can easily be adapted for complex-shaped substrates as well as flat, cylindrical, etc. [11]. This method has been used with numerous materials, for a wide range of applications such as hydroxyapatite for biomedical applications [12], electrolytes for solid oxide fuel cells [13], luminescent materials [14], functionally graded ceramics [15], carbon nanotubes [16], superconductors [17], nanosize zeolite membranes [18], etc. EPD also has industrial and medical applications such as corrosion protection processes.

In this paper, p-type Bi_2Te_3 films has been deposited using the EPD method, and a high quality crack-free thick film has been achieved for TE applications.

II. Experimental Procedure

Bi_2Te_3 powder was commercially available from the Bio-chemical Leader Company with 99.99% purity. The chemical composition of p-type Bi_2Te_3 powder was 22.51 Bi (wt.%), 54.90 Te (wt.%) and 23.59 Sb (wt.%). The median particle size (D50) of 20 µm was characterized for particle sizes of the powder using a Malvern Mastersizer 2000 laser particle size analyser. The SEM image of the starting powder is shown in Fig. 1 at diverse magnifications.

A stable EPD suspension of p-type Bi_2Te_3 was prepared in a mixture of 25% Vol acetone + 75% Vol ethanol with 0.02g triethanolamine as a stabilizer for 100 ml of the media mixture. The purity of the acetone, ethanol, and triethanolamine supplied by the Sigma Aldrich Company was 99.5%, 99.8% and 99.5%, respectively.

Fig. 1 The SEM image of p-type Bi_2Te_3 powder at diverse magnifications (1000x and 5000x)

To prepare the suspension, 1 g of Bi_2Te_3 powders was dispersed in 100 ml of the media mixture (10 g/L Bi_2Te_3 suspension). The mixture was put in an ultrasonic bath for 15 min to disperse and break the agglomerates and stabilise the suspension. The suspension was then rested for at least for 2 hours to reach stability. The suspension was ultrasonicated for 2 minutes prior to the EPD process.

A schematic image of the EPD cell is shown in Fig. 2. The distance between the substrate and the counter electrode was constant at 10 mm for the EPD process. The electrodes were vertical and kept parallel and fixed during the EPD process.

The substrate and the counter electrode were cut from a commercially obtained copper foil roll, with dimensions of 10*10 mm. A range of deposition voltages (0-100 V) and times (1-10 minutes) have been investigated for the EPD process.

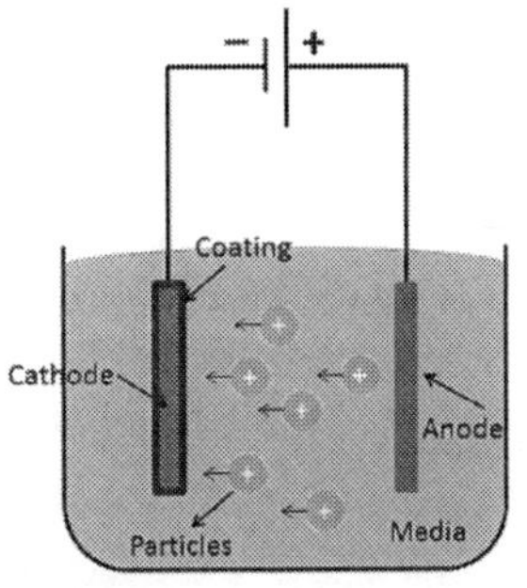

Fig. 2 Schematic image of the EPD cell

III. Results and Discussion

A. The Effect of EPD Time and Voltage on the Deposition Weight

According to Hamaker's equation [19], the EPD deposit yield (w) is calculated through the following equation:

$$w = \int_{t_1}^{t_2} \mu . E . A . C . dt$$

where μ is the electrophoretic mobility; E is the electric field strength; A is the surface area of the electrode; C is the particle mass concentration and t is the deposition time.

Fig. 3 shows the deposition weights of the suspension at different applied voltages and deposition times. It shows that the deposition weights increase when the applied voltages or deposition times increase. It also shows that there is a linear dependence between these two EPD factors and the deposition weights, which is in good agreement with Hamaker's equation and EPD theoretical principles.

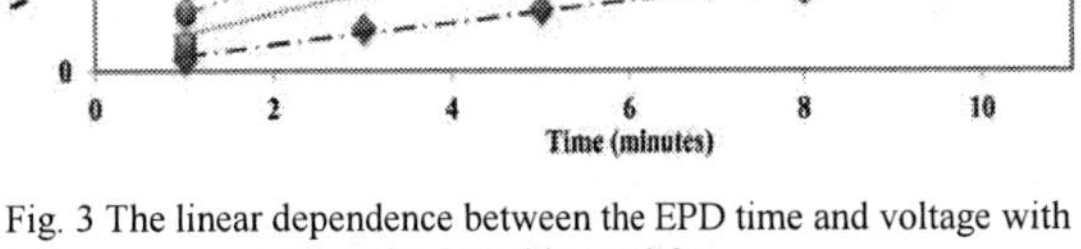

Fig. 3 The linear dependence between the EPD time and voltage with the deposition weight

B. The effect of EPD Voltage on the Film Microstructure

The applied voltage can affect the macro and microstructure of the deposited film. Fig. 4 shows the SEM images of green

Bi_2Te_3 film surface topography at two different EPD voltages of 100 V (a) and 50 V (b) for 10 minutes. It shows that by decreasing the voltage, the roughness of the surface decreases. However, in principal, this can lead to a reduction in the mechanical properties as well as the deposition weight and thickness of the Bi_2Te_3 films, which are desirable neither for handling the green films nor for the potential TE applications [8], [9], [11]. Therefore, in order to accomplish a high quality film, achieving an optimal voltage and time is essential for EPD [8].

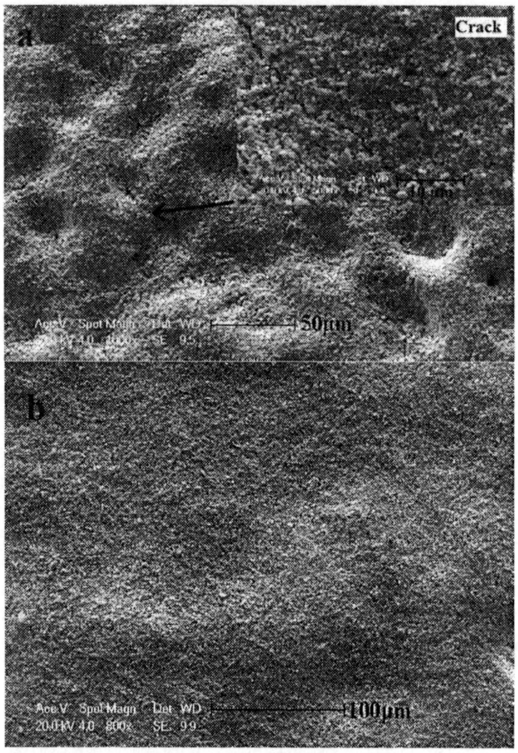

Fig. 4 The SEM images of green Bi_2Te_3 films at two different EPD voltages, (a) 100V and (b) 50V

Although the results showed that the films deposited at 100V had a better interfacial adhesion to the substrate and higher deposition weight, there were small cracks in the film, and the surface was not smooth enough, as can be seen from Fig. 4 (a). Therefore, a voltage of 50 V is considered as the optimum deposition voltage for this suspension because it has deposited a film with higher quality.

IV. Conclusion

In summary, a crack-free and high quality p-type Bi_2Te_3 film has been successfully deposited by EPD for TE applications for the first time. A stable EPD suspension of p-type Bi_2Te_3 has been prepared in a mixture of acetone-ethanol with the use of triethanolamine as a stabilizer. The results have shown that the relationships between the deposition weight and EPD-related process parameters such as applied voltage and time were linear, which is all in good agreement with EPD theoretical principles. The optimum voltage and time of the EPD process was 50 V and 10 minutes, respectively, to achieve a high quality film on a copper substrate. An investigation of the morphology and microstructures of the green deposited films showed that a high quality, crack-free thick film of p-type Bi_2Te_3 has been achieved for TE applications.

Acknowledgment

This work has been supported by the Australian Government Research Training Program Scholarship and The University of Adelaide.

References

[1] Snyder GJ, Toberer ES. Complex thermoelectric materials. Nature Materials. 2008; 7: 105-14.
[2] Shakouri A. Recent developments in semiconductor thermoelectric physics and materials. Annual Review of Materials Research. 2011;41: 399-431.
[3] Dresselhaus MS, Chen G, Tang MY, Yang RG, Lee H, Wang DZ, et al. New directions for low-dimensional thermoelectric materials. Advanced Materials. 2007; 19: 1043-53.
[4] Hamid Elsheikh M, Shnawah DA, Sabri MFM, Said SBM, Haji Hassan M, Ali Bashir MB, et al. A review on thermoelectric renewable energy: Principle parameters that affect their performance. Renewable and Sustainable Energy Reviews. 2014;30:337-55.
[5] Koumoto K, Terasaki I, Funahashi R. Complex oxide materials for potential thermoelectric applications. MRS Bulletin. 2006;31:206-10.
[6] LeBlanc S, Yee SK, Scullin ML, Dames C, Goodson KE. Material and manufacturing cost considerations for thermoelectrics. Renewable and Sustainable Energy Reviews. 2014;32:313-27.
[7] DiSalvo FJ. Thermoelectric cooling and power generation. Science. 1999;285:703-6.
[8] Besra L, Liu M. A review on fundamentals and applications of electrophoretic deposition (EPD). Progress in Materials Science. 2007;52:1-61.
[9] Corni I, Ryan MP, Boccaccini AR. Electrophoretic deposition: From traditional ceramics to nanotechnology. Journal of the European Ceramic Society. 2008;28:1353-67.
[10] Ferrari B, Moreno R. EPD kinetics: A review. Journal of the European Ceramic Society. 2010;30:1069-78.
[11] Boccaccini AR, Zhitomirsky I. Application of electrophoretic and electrolytic deposition techniques in ceramics processing. Current Opinion in Solid State and Materials Science. 2002;6:251-60.
[12] Wei M, Ruys AJ, Milthorpe BK, Sorrell CC, Evans JH. Electrophoretic deposition of hydroxyapatite coatings on metal substrates: A nanoparticulate dual-coating approach. Journal of Sol-Gel Science and Technology. 2001;21:39-48.
[13] Talebi T, Raissi B, Maghsoudipour A. Electrophoretic deposition of YSZ electrolyte on porous NiO-YSZ substrate for solid oxide fuel cells. Key Engineering Materials. 2009;412:215- 20.
[14] Shane MJ, Talbot JB, Kinney BG, Sluzky E, Hesse KR. Electrophoretic deposition of phosphors. Journal of Colloid and Interface Science. 1994;165:334-40.
[15] Put S, Vleugels J, Anné G, Van der Biest O. Functionally graded ceramic and ceramic–metal composites shaped by electrophoretic deposition. Colloids and Surfaces A: Physicochemical and Engineering Aspects. 2003;222:223-32.
[16] Boccaccini AR, Cho J, Subhani T, Kaya C, Kaya F. Electrophoretic deposition of carbon nanotube–ceramic nanocomposites. Journal of the European Ceramic Society. 2010;30:1115-29.
[17] Maiti HS, Datta S, Basu RN. High-Tc Superconductor Coating on Metal Substrates by an Electrophoretic Technique. Journal of the American Ceramic Society. 1989;72:1733-5.
[18] Shan W, Zhang Y, Yang W, Ke C, Gao Z, Ye Y, et al. Electrophoretic deposition of nanosized zeolites in non-aqueous medium and its

application in fabricating thin zeolite membranes. Microporous and Mesoporous Materials. 2004;69:35-42.
[19] Hamaker HC. Formation of a deposit by electrophoresis. Transactions of the Faraday Society. 1940;35:279-87.
[20] Ata MS, Zhitomirsky I. Electrophoretic nanotechnology of ceramic films. Advances in Applied Ceramics: Structural, Functional & Bioceramics. 2012;111:345-50.
[21] Boccaccini AR, Cho J, Roether JA, Thomas BJC, Jane Minay E, Shaffer MSP. Electrophoretic deposition of carbon nanotubes. Carbon. 2006;44:3149-60.
[22] Van Tassel JJ, Randall CA. Ionic Gradients at an Electrode above the Equilibrium Limit Current. 1. Concentration and Charge Gradients in an Immobile Electrolyte. The Journal of Physical Chemistry C. 2007;111:3341-8.
[23] Van Tassel JJ, Randall CA. Role of ion depletion in the electrophoretic deposition of alumina powder from ethanol with increasing quantities of HCl. Journal of Materials Science. 2006;41:8031-46.
[24] Nobuyuki K, Takeyo T, Hiromasa S, Touru H. Preparation of Various Oxide Films by an Electrophoretic Deposition Method: A Study of the Mechanism. Japanese Journal of Applied Physics. 1995;34:1643.

Distributed Automation System Based Remote Monitoring of Power Quality Disturbance on LV Network

Emmanuel D. Buedi, K. O. Boateng, Griffith S. Klogo

***Abstract*—**Electrical distribution networks are prone to power quality disturbances originating from the complexity of the distribution network, mode of distribution (overhead or underground) and types of loads used by customers. Data on the types of disturbances present and frequency of occurrence is needed for economic evaluation and hence finding solution to the problem. Utility companies have resorted to using secondary power quality devices such as smart meters to help gather the required data. Even though this approach is easier to adopt, data gathered from these devices may not serve the required purpose, since the installation of these devices in the electrical network usually does not conform to available PQM placement methods. This paper presents a design of a PQM that is capable of integrating into an existing DAS infrastructure to take advantage of available placement methodologies. The monitoring component of the design is implemented and installed to monitor an existing LV network. Data from the monitor is analyzed and presented. A portion of the LV network of the Electricity Company of Ghana is modeled in MATLAB-Simulink and analyzed under various earth fault conditions. The results presented show the ability of the PQM to detect and analyze PQ disturbance such as voltage sag and overvoltage. By adopting a placement methodology and installing these nodes, utilities are assured of accurate and reliable information with respect to the quality of power delivered to consumers.

***Keywords*—**Power quality, remote monitoring, distributed automation system, economic evaluation, LV network.

I. Introduction

THE current technological advancement and sophistication has resulted in the computerization of domestic and industrial equipment, and hence, an increase in their sensitivity to the quality of power they depend on. Utility companies the world over have an additional mandate to ensure that the power they supply to consumers is free from disturbances. Total elimination of these disturbances from the distribution network has proven to be impossible. This is due to the complexity of the network and mode of operation of equipment connected to the network.

In order to resolve power quality (PQ) disturbances, utility companies rely on power quality monitors (PQMs) to detect and hence find solutions to the problem. To perform a complete evaluation of power quality disturbance in the network, the steps outlined in Fig. 1 can serve as a guide [1]-[5].

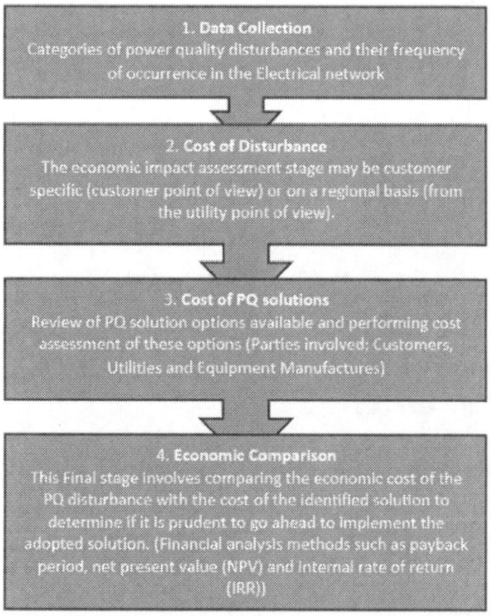

Fig. 1 Steps involved in performing a complete economic evaluation of power quality disturbance

The first step, which involves the collection of data on various forms of power quality disturbances and their frequency of occurrence, is very vital. Any error in this process affects the whole evaluation process.

The type of monitors and their mode of operation will have an impact on the data collected. Some power measurement and monitoring equipment such as energy meters have been equipped with secondary PQM capabilities.

The conventional method of placement of these energy meters is to install them on every feeder to check revenue losses in the network. References [6], [12], [13] show how network automation equipment such as RTU and energy meters can be upgraded to perform power quality monitoring. This may seem as a cheaper alternative for utility companies

Emmanuel D. Buedi is with the Electricity Company of Ghana, Supervisory Systems Division, P.O.Box, 521, Accra, Ghana (phone: +233266188177, e-mail: inteldana09@gmail.com).

K. O. Boateng is with the Kwame Nkrumah University of Science and Technology (KNUST) Department of Computer Engineering, PMB, UPO, KNUST, Kumasi, Ghana (phone: +233203154862, e-mail: koboat_2000@yahoo.com).

Griffith S. Klogo is with the KNUST Department of Computer Engineering, PMB, UPO, KNUST, Kumasi, Ghana (phone: +233244880568, e-mail: klogoselorm@yahoo.com).

but will end up affecting data collected. Factors affecting the decision and placement of this network equipment are mostly conventional and differ from placement methods for PQMs. This may end up increasing redundant data or reducing network coverage of the monitor. The availability of new monitor placement algorithms has proven to be efficient in using the minimum amount of monitors to provide a better network observability and reduced data redundancy at less cost [8]-[10]. There is therefore the need to focus on devices specifically designed and dedicated for monitoring PQ disturbances.

Power quality monitors and analyzers could come in the form of handheld, standalone or networked. The choice of a PQM to use depends on a number of considerations such as the type of disturbance and the purpose of measurement. Measurement equipment ranging from simple voltage or current meters to highly sophisticated analyzers and monitors are available. Fig. 2 categorizes this equipment depending on their mode of operation. Primary PQMs refer to equipment used purposely to measure or sample voltage or current parameters for analysis. This category of devices can be temporally or permanently installed in the electrical distribution network. Temporal PQMs refer mostly to handheld or easily movable devices used to measure or monitor disturbances in the network. They range from simple voltmeters, ammeters, oscilloscopes to sophisticated handheld analyzers such as the Fluke 435 series and AEMC 8435 PowerPad. These devices can be installed over a period and their recorded data can be viewed onscreen or downloaded onto a PC. Permanent PQMs on the other hand refers to the monitors that are permanently installed in the network. They are designed to operate autonomously with weather and power considerations in mind. They could be standalone or be monitored remotely over a communication network. Secondary PQMs may refer to any equipment designed to have additional features to measure voltage and current parameters for analysis. Devices that fall under this category include some (smart) revenue meters, IEDs and RTUs or FTUs. Primarily, revenue meters are designed to measure the amount of power consumed in KWh for billing purposes. RTU/FTU is used to monitor and control field devices on the distribution network. Since these devices are installed in the network, some manufactures have decided to include disturbance monitoring functionalities due to its importance to utility companies and consumers. Since power quality monitoring is not their primary function, the decision as to where to install these devices in the network may render the PQ data recorded less relevant.

Individual consumers or industries can use handheld or standalone monitors at their premises, but for utility companies, the networked monitors are preferred. Most utility companies can make use of their existing DAS infrastructure to provide connectivity to the PQMs. Data collected from the monitors is centralized making analysis faster and easier.

This study presents an overview of a dedicated PQM capable of integrating into an existing DAS infrastructure. This work focuses more on voltage sags since it has been identified as the most important power quality problems affecting industries and large commercial customers [1]. A portion of an existing network is modeled and analyzed using the SimPowerSystem toolbox in MATLAB. The monitoring unit of the designed PQM is implemented and installed in an LV network over a period of time.

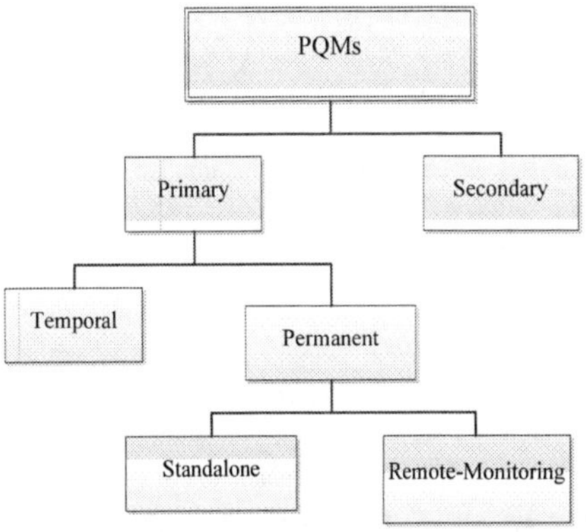

Fig. 2 Categorization of power quality monitors

This paper is organized as follows. Section II discusses the various PQ disturbances available in the electrical network. The overview of the designed PQM is presented in Section III. In Section IV, the model of a portion of the network is presented and the outcome of the simulation is analyzed. Finally, real-time data gathered from the PQM installed in the network is presented in Section V.

II. Power Quality Disturbances

A number of problems regarding the quality of power delivered by utility companies can be identified [2], [4], [5]. Voltage sags/dip, voltage swells, under voltage, interruption, transients, voltage unbalance, voltage fluctuations and harmonics are some major power quality problems in a distribution network. These system disturbances are categorized by factors such as magnitude of disturbance and duration. Among these disturbances, voltage sag and momentary interruption have been identified as the most important in terms of their impact on equipment operations [1], [7], [11].

A. Voltage Sag

Voltage sags refers to the reduction in magnitude of voltage level between 10% to 90% of the RMS and last for duration of 8 milliseconds to 60 seconds. The causes of voltage sags can be categorized into two main types: consumer side and utility side.

At the consumer side, voltage sags can arise during the startup of large loads such as industrial motors or by electrical faults. There is a significant instantaneous drop in voltage level due to the large initial current drawn by these loads.

At the utility provider's end, causes of voltage sags includes: operation of reclosers and circuit breakers, equipment failure due to overloading or cable faults, thunder storms or lightening, salt spray that builds up around conductors close to coastal areas and animals or birds coming into contact with electrical distribution equipment.

B. Overvoltage

Overvoltage refers to an increase in the AC supply voltage in the electrical distribution network for a period of time. Depending on the magnitude of increase and duration, overvoltage can be classified into two main categories.

Momentary overvoltage or voltage swell refers to RMS variation that exceeds 110% of the nominal voltage and last for the duration of 0.5 cycles to 60 seconds. This is usually caused by faults or brief interruption of power lines. Long duration overvoltage on the other hand refers to RMS variations that exceeds 110% of the nominal voltage and lasts for more than 60seconds.

They are usually caused by wrong transformer tap settings, decrease in load on a phase, lightening and switching of a capacitor bank connected to the network.

C. Interruption

Interruption refers to a complete loss of voltage or load current in one or more phases of a source. Depending on the duration, interruption can be grouped into three main categories. These include momentary interruptions, temporal interruptions and sustained interruptions. Momentary interruption refers to loss of voltage for a period of between 8 milliseconds and 3 seconds. Temporal interruption refers to a drop of voltage below 10% of the nominal voltage level (i.e. < 23VAC) for a period of between 3 seconds and 60 seconds. Sustained interruption refers to continues loss of voltage for more than 60 seconds in an electrical system. The economic impact of these interruptions ranges from permanent loss of data on a computer to massive damage of industrial equipment. Three seconds of power interruption can cause damage that will require more than 7 hours to repair.

D. Transient

Transient refers to a sudden increase or decrease of electrical current or voltage that dissipates in a short duration. Impulsive transient refers to a sudden increase or rise in voltage or current in an electrical system in the negative or positive direction. The sudden rise to a peak and decay to nominal value takes place in less than 50 nanoseconds. Lightening, electrostatic discharge and switching of inductive loads are some of the identifiable causes of impulsive transients or "spikes". Mounting of lightening arresters and proper grounding of transformers and other electrical distribution equipment are some solutions to this problem. Oscillatory transient do not decay quickly, but rather oscillate for 0.5 to 3 cycles at the nominal frequency.

III. Design of PQM with Remote Monitoring Function

Data gathered from PQMs serves as a vital source of information during decision making in the quest for a solution to disturbances in the network.

The main aim of remote monitoring is to ensure that real-time data gathered is made available instantly for analysis by system operators. Fig. 3 gives a design of a PQM with remote monitoring features.

The system consists of PQMs installed at monitoring points on the network and sending data to a centralized location. This follows a typical distributed automation system (DAS) architecture, which is already common in the electrical distribution sector. Hence, this system can easily be integrated into an existing AMR or SCADA system.

The monitor consists of the power supply unit, the control unit and the sensor board. The power supply unit ensures that the system gets the right voltage level for its operation. It also has battery backup to keep the system running in case of a long duration interruption. The sensor board consists of current and voltage sensors to pick analog data from the monitoring point in the network for the control unit.

The control logic board also makes use of the ADE7558 IC, which is a high accuracy 3-phase electrical energy measurement IC with a serial interface and two pulse outputs. This IC can measure RMS voltage and current, active/reactive/ apparent energy, frequency and sampled waveform data; it also contains a circuitry for overvoltage and voltage sags with programmable thresholds. When these thresholds are exceeded, the IC generates a fault alert, which serves as an interrupt for the microcontroller. Fig. 4 shows the flowchart of the system.

The program starts by initializing the ADE7758 IC and the second memory unit (Mem 2). This process calibrates the ADE7558 IC by setting the thresholds for voltage sags and overvoltage. The ADC attached to the controller constantly monitors and digitizes the AC signal into Mem 1. This data is transmitted intermittently to the server. The ADE7758 IC also monitors the AC signal for any abnormalities. If any fault is detected, the controller is alerted and the digitized faulty signal is transferred into Mem 2 for immediate transmission. The memory units are cleared after every data transmission.

These units communicate with the control center the same way as RTUs and AMR meters. The servers at the monitoring station intermittently polls the units to read information. The units are also set to transmit immediately new and urgent information to the servers.

Fig. 5 shows a typical communication network overview of the setup for the control center and PQMs installed in the electrical network. To maintain uniformity and conform to industrial standards, the data sampled by the PQM is formatted into an IEC 60870-5-101 frame, as shown in Fig. 6, before transmission to the server.

IV. Model of the Network

A portion of the distribution network of the Electricity Company of Ghana showing the power supply from the substation to the end user is shown in Fig. 7. The standard voltage levels in the network and the PQ monitoring points are indicated in this diagram.

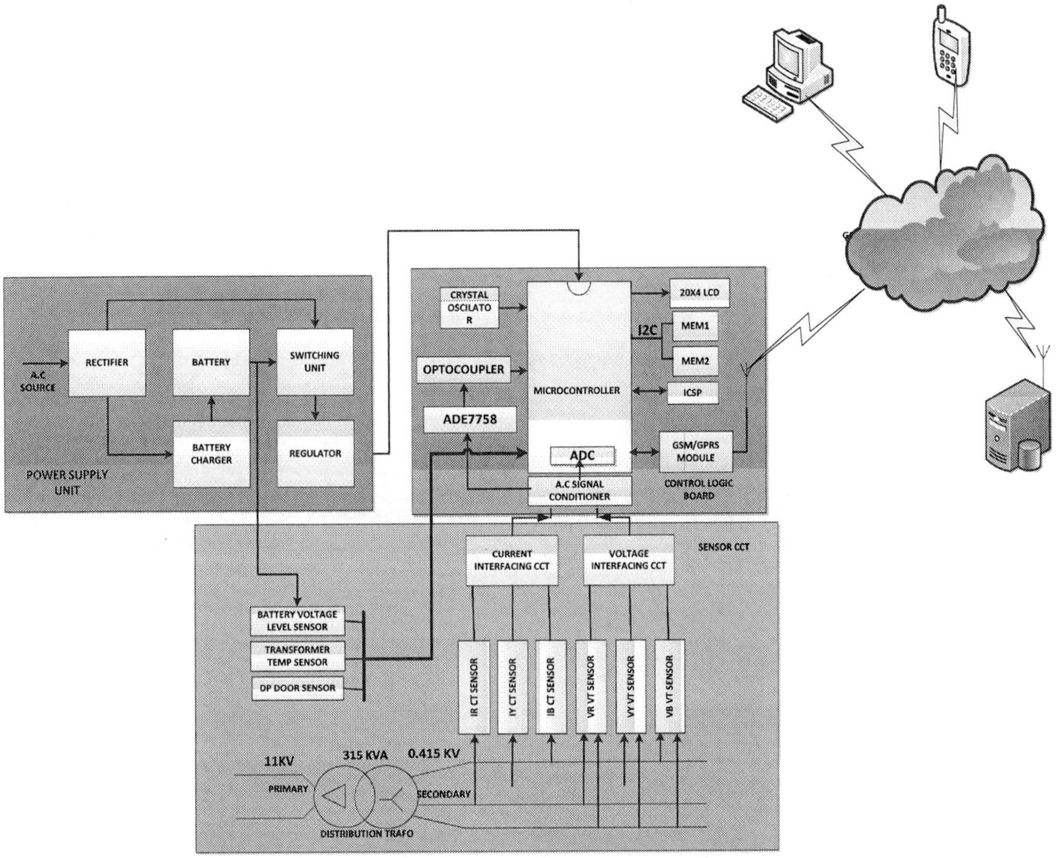

Fig. 3 Design of PQM

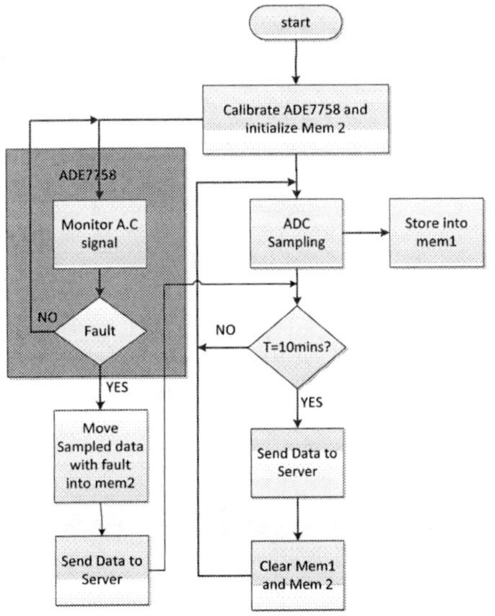

Fig. 4 Flow chart for controller

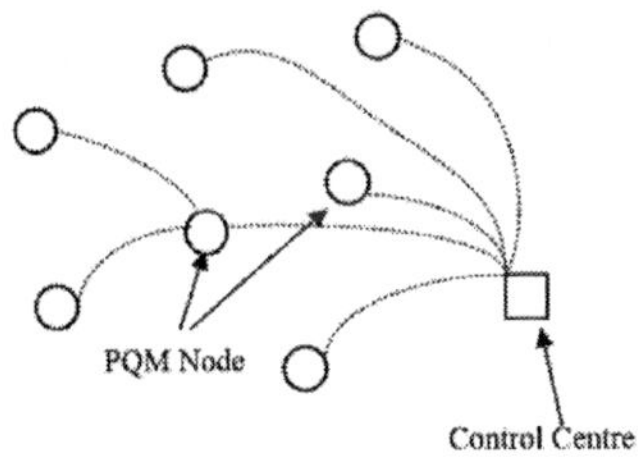

Fig. 5 PQM network setup

The SimPowerSystem library in Simulink provides a comprehensive set of models for the components required in this work and has been used to generate a model for the circuit given in Fig. 7. The Three-Phase source block implements a balanced three-phase voltage source with internal R-L impedance. The source used in this work has 11000VAC phase-to-phase RMS, frequency of 50Hz, source impedance of 0.8929 and inductance of 16.58e-3. Three-Phase Breaker block implements a three-phase circuit breaker where the opening and closing times can be controlled either from an external Simulink signal (external control mode), or from an internal control timer (internal control mode). The Three-

Phase Transformer (Two Windings) block implements a three-phase transformer using three single-phase transformers which steps down the voltage from the source to 415VAC. The Three-Phase Fault block uses three Breaker blocks that can be individually switched on and off to program phase-to-phase faults, phase-to-ground faults, or a combination of phase-to-phase and ground faults. The time for the fault to occur can be controlled internally or externally. The customer facility is represented by a series RLC branch. The Series RLC Branch block implements a single resistor, inductor, or capacitor, or a series combination of these. The values of each parameter can be set to give a representation of the customer's load characteristics. The voltages and currents at various stages in the circuit are monitored by the Three-Phase V-I Measurement block and the scopes. Results of simulation carried out on the model presented in Fig. 8 are presented in this section. The Fault block is set to introduce an earth-fault into the network from t=0.1 to t=0.2 during the simulation. The results of an earth fault on the yellow phase are presented in Fig. 9. This leads to a momentary interruption on the yellow phase and voltage increase on the other phases. Fig. 10 presents a two-phase bridging fault between the yellow and blue phases. This fault causes a voltage sag on both phases and the red phase is not affected.

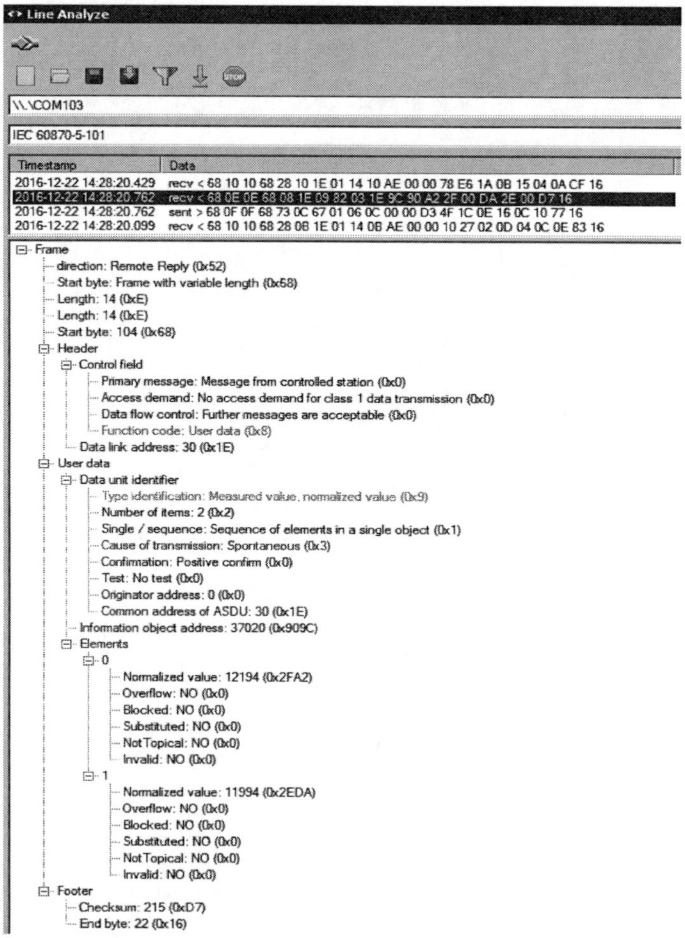

Fig. 6 Format of sampled analog data sent to server

Fig. 11 shows the behavior of the model output under different sampling rates. From this figure, it can be observed that the signal starts to deviate considerably for rates below 0.5 kHz. Any rate above this threshold is enough to represent accurately the signal for analysis.

The RMS profile for an AC signal is a preferable way to represent signals for analysis and comparison. The logic of the PQM is then set to report any RMS value below 0.9 pu as voltage sag. The output RMS profile of the PQ disturbance from the LV network is given in Fig. 12. This shows how the PQM interprets a voltage sag event.

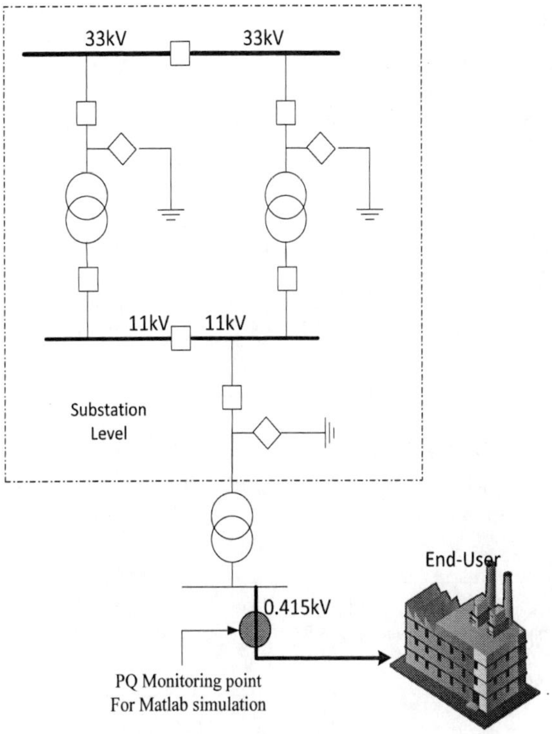

Fig. 7 Single Line Diagram describing energy supply from substation to End-user facility

V. Results and Discussion

To assess the performance of the designed PQM in a real environment, the monitoring unit of the PQM is implemented and subjected to data with PQ disturbances from an existing LV network, as shown in Fig. 14. A normal single-phase voltage signal captured from the network is presented in Fig. 13.

With a sampling rate of approximately 2 kHz, this monitoring unit is able to capture the voltage waveform in good detail for analysis. Data presented in Fig. 15 shows the various instantaneous voltage levels sampled and corresponding RMS calculated with a window of 10 units. The corresponding graphical output of this data is presented in Fig. 14.

VI. Conclusion

This paper has identified and presented the reasons for utilities to invest into DAS-based PQMs as opposed to secondary PQMs. A proposed design of this recommended type of PQM has been presented and the monitoring unit of this design has been implemented. This work also presents the Simulink module of a portion of an existing network and simulates various short circuit faults. Results in both cases show the ability of the system to detect and analyze disturbance in the signal waveform. Utilities can deploy these PQM nodes in their network based on a chosen optimal placement algorithm and hook it onto an existing SCADA or AMR communication infrastructure. Data received from these nodes can be used as inputs for economic evaluation of power quality disturbance in the network.

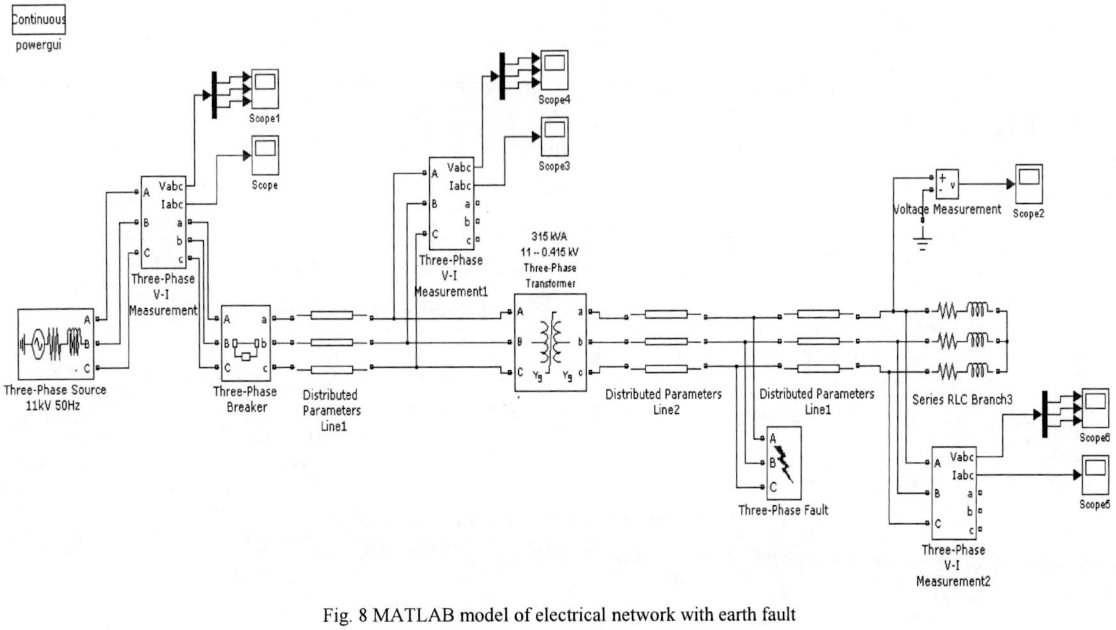

Fig. 8 MATLAB model of electrical network with earth fault

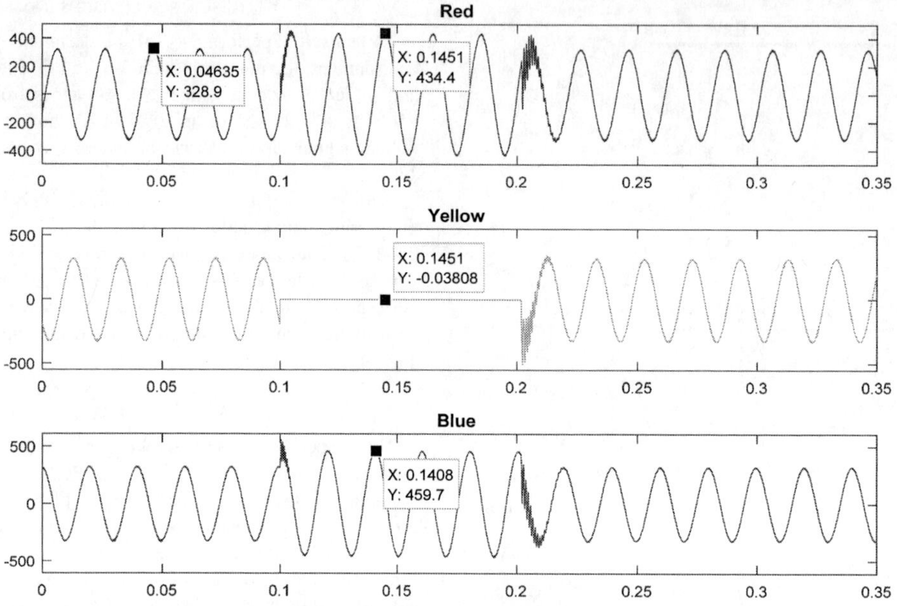

Fig. 9 Earth Fault on the yellow phase for Vpeak value of 328.9

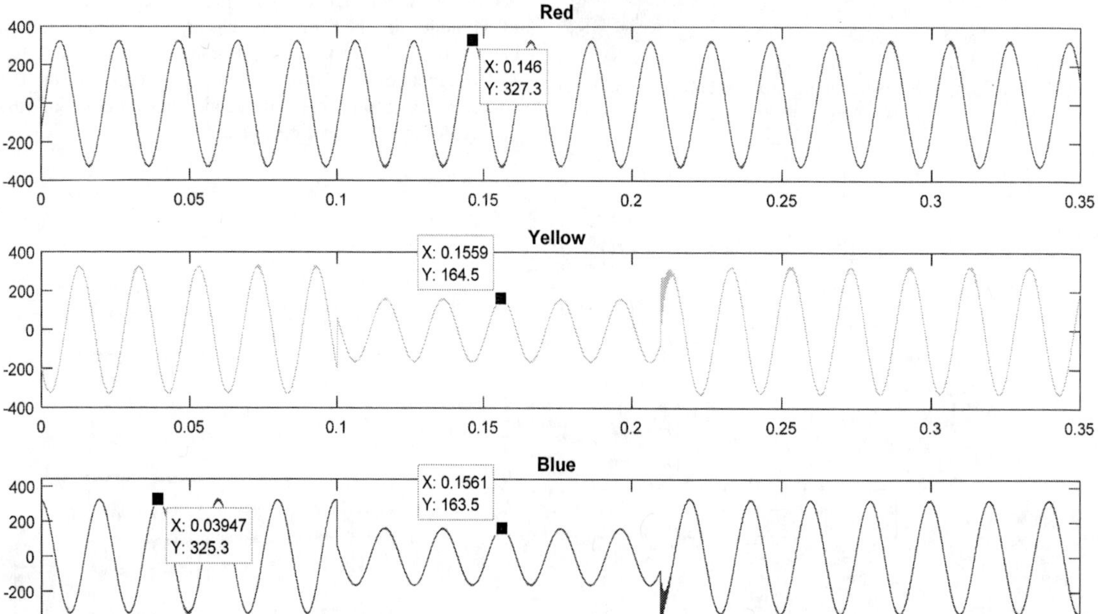

Fig. 10 Short circuit fault between yellow and blue phase

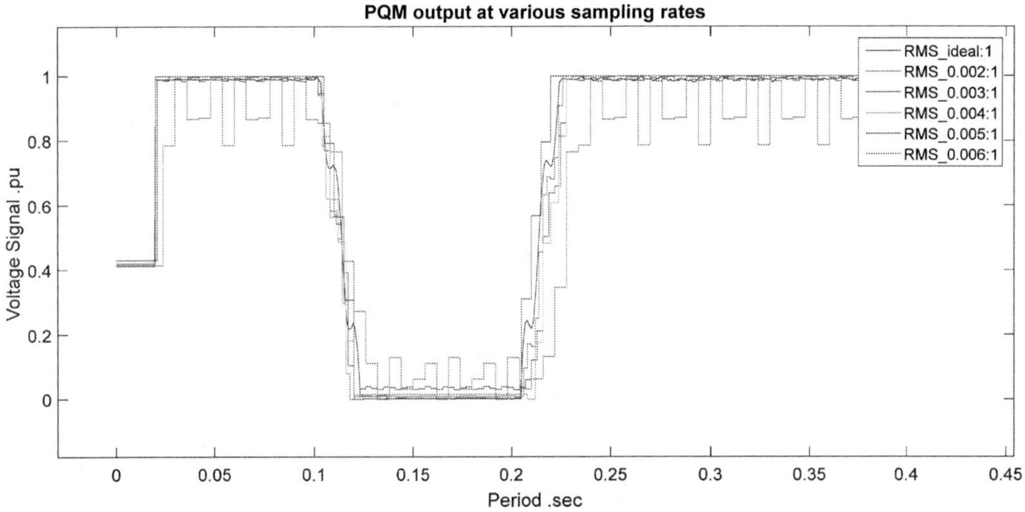

Fig. 11 PQM outputs at various sampling rates

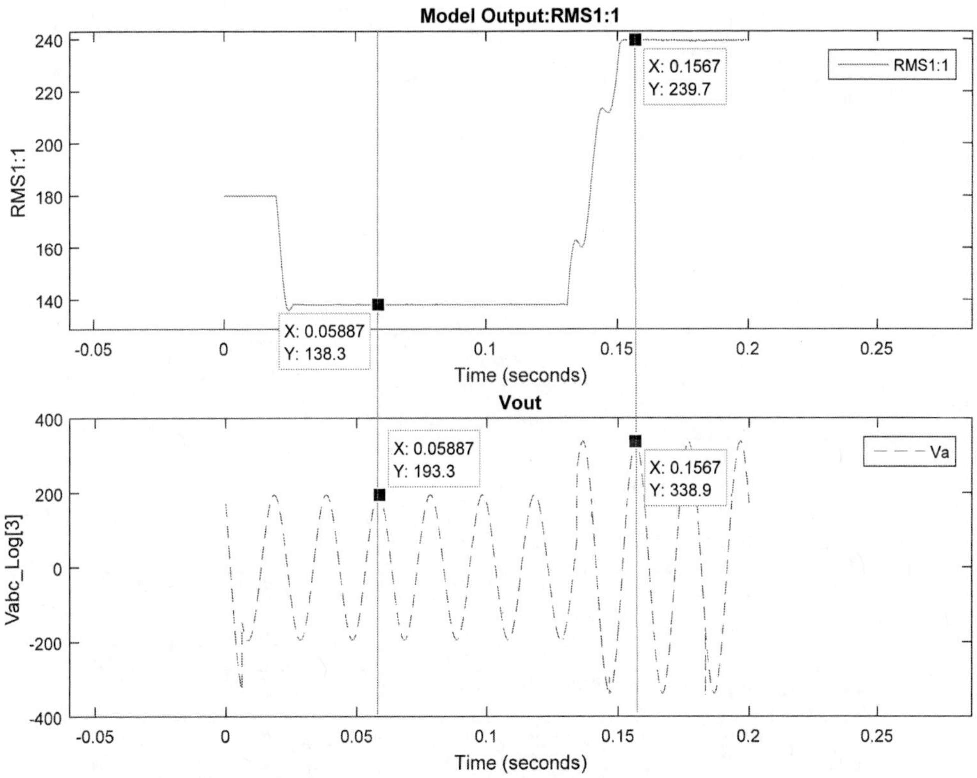

Fig. 12 PQM output RMS profile for voltage sag

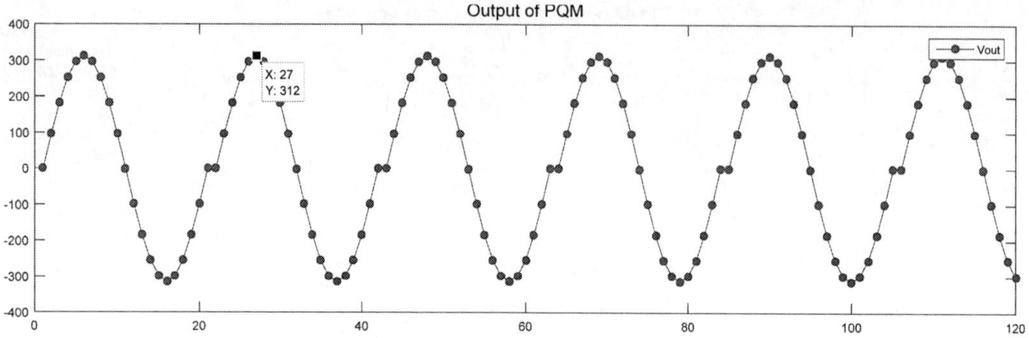

Fig. 13 PQM output for real-time voltage signal from an existing single-phase LV network

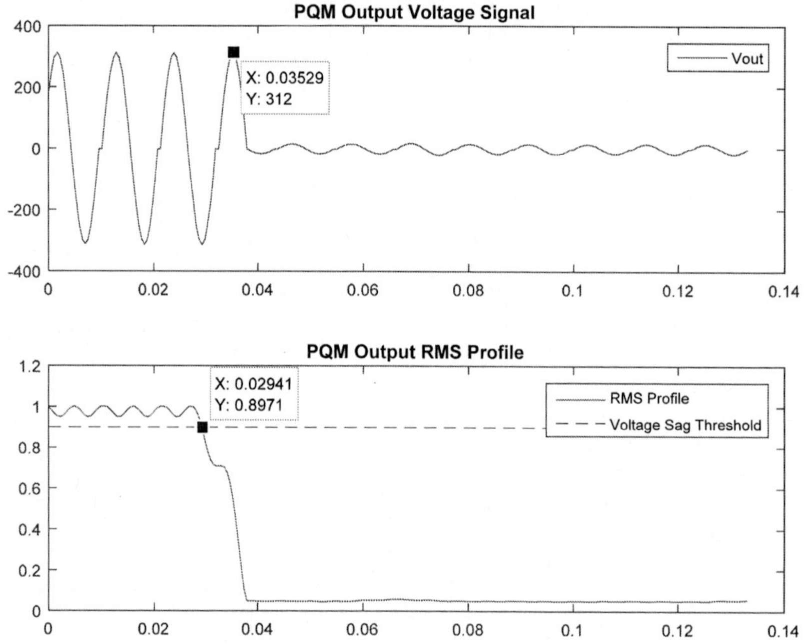

Fig. 14 Voltage sag disturbance captured by PQM

ts	0.030481	0.031016	0.031551	0.032086	0.03262	0.033155	0.03369	0.034225	0.034759	0.035294
Vout	-252.7	-183.7	-96.4	0.0	0.0	95.4	181.4	249.4	299.7	312.0
Vrms	0.783409	0.740363	0.71663	0.70996	0.709977	0.709977	0.703181	0.678045	0.627705	0.550653
ts	0.035829	0.036364	0.036898	0.037433	0.037968	0.038503	0.039037	0.039572	0.040107	0.040642
Vout	296.7	251.9	178.4	98.4	0.0	-4.4	-9.8	-12.6	-14.8	-15.6
Vrms	0.450426	0.334927	0.215835	0.109572	0.0499	0.05014	0.0499	0.049267	0.048472	0.047819
ts	0.041176	0.041711	0.042246	0.042781	0.043316	0.04385	0.044385	0.04492	0.045455	0.045989
Vout	-14.1	-12.6	-9.2	-4.8	0.0	0.0	4.8	9.2	12.6	14.8
Vrms	0.047567	0.047819	0.048472	0.049267	0.0499	0.05014	0.05014	0.0499	0.049267	0.048472
ts	0.046524	0.047059	0.047594	0.048128	0.048663	0.049198	0.049733	0.050267		
Vout	15.6	14.8	12.6	9.7	4.1	0.0	-4.8	-9.2		
Vrms	0.047819	0.047567	0.047819	0.048472	0.049267	0.0499	0.05014	0.0499		

Fig. 15 Data from the LV network

REFERENCES

[1] Mark McGranaghan, Bill Roettger (2002, Feb). Economic Evaluation of Power Quality. *IEEE Power Engineering Review* (Online). 22(2), pp. 8-12. (Accessed: 25- June- 2015). Available: http://ieeexplore.ieee.org/xpl/abstractMetrics.jsp?arnumber=981339&ranges=2002_2003_Year&searchWithin=%22Authors%22:.QT.M.%20McGranaghan.QT

[2] IEEE Recommended Practice for Evaluating Electric Power System Compatibility With Electronic Process Equipment, IEEE Standard 1346, 1998.

[3] D. coll-mayor, j. Pardo, M. Perez-Donsion (2009, April). Analysis of Methodologies for the Economical Evaluation of Power Quality. International Conference on Renewable Energies and Power Quality (ICREPQ'09).

[4] Barry W. Kennedy, "Power Quality Economics," in Power Quality Primer, New York: McGraw-Hill, 2000, pp. 239–271.

[5] Heine, P., Pohjanheimo, P., Lehtonen, M., Lakervi, E (2002, May). A Method for Estimating the Frequency and Cost of Voltage Sags, *IEEE Transactions on Power Systems*, 17(2), pp. 290-296.

[6] Boknam Ha, Shinyeol Park, Changhoon Shin, Seongchul Kwon, Soyeong Park, (2007, May). Power Quality Monitoring On Distribution Networks Using Distribution Automation System, *19th International Conference on Electricity Distribution (C I R E D '09),* Paper 0426.

[7] Math H. J. Bollen (2000, October). *Understanding power quality problems: voltage sags and interruptions.* (Online). (Accessed: 25- June-2015). Available: http://www.wiley.com/WileyCDA/WileyTitle/productCd-0780347137.html

[8] Ahmad Asrul Ibrahim, Azah Mohamed, Hussain Shareef, Sakti Prasad Ghoshal (2012, March). Optimization Methods for Optimal Power Quality Monitor Placement in Power Systems: A Performance Comparison.International Journal on Electrical Engineering and Informa tics - Volume 4(1).

[9] Asadollah Kazemi, Azah Mohamed, Hussain Shareef, Hadi Zayandehroodi (2013). A Review of Power Quality Monitor Placement Methods in Transmission and Distribution Systems. Przegląd Elektrotechniczny, ISSN 0033-2097, R. 89 NR 185-188.

[10] Ibrahim, A.A., A. Mohamed, H. Shareef. (2010). Optimal Placement of Voltage Sag Monitors Based on Monitor Reach Area and Sag Severity Index. IEEE Student Conference on Research and Development (SCOReD) pp.467-470.

[11] Jeff Lamoree, Dave Mueller, Paul Vinett, William Jones (1994). Voltage Sag Analysis Case Studies. IEEE Transactions on Industry Applications 30(4) pp.1083-1089.

[12] Michael Campbell, Neville Watson, Allan Miller (2015). Smart Meters to Monitor Power Quality at Consumer Premises. EEA Conference & Exhibition.

[13] Shihui Yang, Fangfang Duan, Ke Li, Bin Li (2015). The Design of Smart Meter with Power Quality Monitoring. International Conference on Applied Science and Engineering Innovation (ASEI) pp1447-1450.

Preparing Data for Calibration of Mechanistic-Empirical Pavement Design Guide in Central Saudi Arabia

Abdulraaof H. Alqaili, Hamad A. Alsoliman

Abstract**—**Through progress in pavement design developments, a pavement design method was developed, which is titled the Mechanistic Empirical Pavement Design Guide (MEPDG). Nowadays, the evolution in roads network and highways is observed in Saudi Arabia as a result of increasing in traffic volume. Therefore, the MEPDG currently is implemented for flexible pavement design by the Saudi Ministry of Transportation. Implementation of MEPDG for local pavement design requires the calibration of distress models under the local conditions (traffic, climate, and materials). This paper aims to prepare data for calibration of MEPDG in Central Saudi Arabia. Thus, the first goal is data collection for the design of flexible pavement from the local conditions of the Riyadh region. Since, the modifying of collected data to input data is needed; the main goal of this paper is the analysis of collected data. The data analysis in this paper includes processing each: Trucks Classification, Traffic Growth Factor, Annual Average Daily Truck Traffic (AADTT), Monthly Adjustment Factors (MAFi), Vehicle Class Distribution (VCD), Truck Hourly Distribution Factors, Axle Load Distribution Factors (ALDF), Number of axle types (single, tandem, and tridem) per truck class, cloud cover percent, and road sections selected for the local calibration. Detailed descriptions of input parameters are explained in this paper, which leads to providing of an approach for successful implementation of MEPDG. Local calibration of MEPDG to the conditions of Riyadh region can be performed based on the findings in this paper.

Keywords**—**Mechanistic-empirical pavement design guide, traffic characteristics, materials properties, climate, Riyadh.

I. Introduction

THE American Association of State Highway and Transportation Officials AASHTO (1961-1993) developed empirical design method for designing pavement structures. This empirical design method has a number of limitations because it is based on performance measurements in one climatic zone, one subgrade type, over two years of testing, and under slightly two million axle load applications [1]. Due to these limitations, the AASHTO Joint Task Force on Pavements (JTFP) updated the empirical method to the mechanistic-empirical method to develop a more accurate design guide. The NCHRP and the AASHTO developed the MEPDG and its software under NCHRP Project 1-37A. The MEPDG program has two modules (mechanistic and empirical). The mechanistic module in the MEPDG program calculates the stresses, strains, and deflection at critical locations in a pavement structure for various loads, materials, and environmental inputs. The empirical module (distress models, also known as the performance prediction models) computes predicted physically distresses such as roughness (International Roughness Index - IRI), rutting, and cracking based on the stresses and strains as well as traffic, materials and environmental inputs. Therefore, extensive and more detailed data are needed for input data in MEPDG. Quantities of data required are vast when comparing with the previous conventional design method (AASHTO 93). Furthermore, to improve the design accuracy, the MEPDG considers the input parameters which effect on the pavement performance including traffic characteristics, environmental condition, and materials properties, as well as the MEPDG approach is a flexible method which allows using local information such as traffic, climate, and materials data [2].

In Saudi Arabia (especially the capital of Riyadh), the recent economic revival has led to an increase in the traffic flow, which in turn, requires the development of a road network. Therefore, The Saudi Ministry of Transportation is currently in the process of implementing the MEPDG for the design of a flexible pavement. The distress models were calibrated using data related to several conditions from across the United States. Calibrating distress models under the local conditions is necessary in order to implement MEPDG for local pavement design. Local data collection is considered the first stage in the MEPDG implementation approach for pavement design. The collected data are used as input data in MEPDG for design procedures and the support process of local calibration together. This paper explains the executed work to obtain local data from the Riyadh region to be used as input data in the MEPDG. Almost of the collected data are needed to be processed to be suitable for input data in MEPDG, and thus, this paper also presents the procedures of collected data modifying. Like Saudi Arabia, the MEPDG has been implemented in many countries. Delgadillo et al. [3] collected local data (such as axle load distributions, weather, and material characteristics) for input data and calibration of MEPDG in Chile. The local traffic data for the design was collected, local weather was used, and material properties were obtained for implementation in MEPDG in India [4]. Zhao et al. [5] determined the axle load distribution factors and number of axles per truck for input traffic data of MEPDG in China. For Costa Rica, Loria et al. [6] presented Costa Rica's experience for the materials properties which are used

Abdulraaof H. Alqaili is a master student at King Saud University, Riyadh, KSA (phone: +966-562101273; e-mail: aalqaili@ksu.edu.sa).
Hamad A. Alsoliman is assistant professor at King Saud University, Riyadh, KSA (e-mail: hasoliman@hotmail.com).

in the MEPDG calibration. Traffic, climate, and materials local data were evaluated and prepared by Caliendo [7] to be used as input data in MEPDG for Flexible Pavement Design in Italy. Depending on the Qatar Highway Design Manual, the materials properties were selected, local traffic loading and environmental conditions were set by Sadek et al. [8] for MEPDG in Qatar.

II. Objectives

The main purpose of this paper is to present a framework to implement the MEPDG for flexible pavements of the Riyadh region and to improve the accuracy of MEPDG pavement performance predictions.

In order to accomplish the main purpose of this paper, the specific objectives are:

- Collecting local data of the flexible pavement such as traffic, environment conditions, materials properties, and pavement performance data from local sources in the Riyadh region.
- Preparing collected data to be suitable as input data for MEPDG and support process of local calibration together.

III. Collection of Local Data

The data collection represents a great part of the study since it plays a significant role in the accuracy of the calibration process. They also require high effort and time because they should be collected from various sources. The required data were collected from the Saudi Ministry of Transportation and Presidency of Meteorology and Environment in Riyadh. In order to use the input parameters that reflected local Riyadh conditions as much as possible, this paper focused on main roads in the Riyadh region which are Riyadh-Wadi Duwaiser, Riyadh-Taif, Riyadh-Qassim, and Riyadh-Dammam. These roads join the Riyadh region with other regions and have the biggest traffic volume, and therefore, they are important.

A. Traffic Data

The MEPDG requires four basic categories of traffic data for the structural pavement design.

1) Traffic volume of base year. AADTT of Vehicle Classes four through 13 is essential data in this category. This information can be derived from vehicle count data.
2) The AADTT of base year should be modified by using adjustment factors such as monthly, hourly, and vehicle class distribution factors. These factors can be calculated from data which is obtained from vehicle count data.
3) The axle load distribution factors (also called axle load spectra) can be determined only from weight stations data.
4) General traffic inputs including number of axles per truck, axle configuration, and wheelbase [9].

The Saudi Ministry of Transportation has advanced electronic devices for counts the traffic volume and classifies the vehicles which contain of Piezoelectric and Scanning Sensors. This process is performed in the permanent traffic counting stations which are located along the main road in the Riyadh region.

The General Directorate of Operation and Maintenance in the Saudi Ministry of Transportation is operating these stations and collects traffic data as a database for the pavement management system. In this paper, traffic data was obtained from the General Directorate of Operation and Maintenance in the Saudi Ministry of Transportation.

The collected data from the permanent traffic counting stations contains the Average Annual Daily Traffic (AADT), hourly distribution traffic volume, percentage of vehicle types, number of lanes, and vehicle classification.

The Average Annual Daily Traffic (AADT) in the year 2012 and percentage of trucks of each road are given in Table I. The number of lanes was three of each road. The dates of the construction of each road (traffic opening) were obtained from the Execution Department in the Saudi Ministry of Transportation as shown in Table I.

TABLE I
AADT, Percent of Trucks, and Dates of Construction

No.	Road Name	AADT (2012) (veh/d) 2 Directions	Percent trucks (%)	Date of construction
10	Riyadh-Wadi Duwaiser	20745	29	1981
40	Riyadh- Taif	30510	31	1981
65	Riyadh- Qassim	22245	22	1982
40	Riyadh- Dammam	19350	44	1981

Truck weigh stations were designed to record the weight of axles and gross weight of the moving vehicle (especially trucks). Truck weigh station provide extensive traffic data which include truck classification, the number of axles, and weight of the axles, distances between the axles, truck length, and gross weight of the truck. Truck weigh stations consist of bending plates, hydraulic load cells, and piezoceramic cables [2]. The General Directorate of Operations and Maintenance in the Saudi Ministry of Transportation operates the truck weigh stations, and takes data from these stations and documents the data to use in pavement management.

In this paper, truck weigh data were obtained from the General Directorate of Operations and Maintenance. These data include the trucks classification, truck length, number of axles, and weight of each axle, as well as the axle distances, and the gross weight of the trucks.

Al-Mansour et al. [10] conducted the tire pressure measurements of the trucks using a pressure gauge with a scale ranging from 0 to 160 psi. Reference [10] shows the distribution of tire pressure values of all trucks and indicates that the range between 120-130 psi was highest. Therefore, tire pressure of 125 psi is used.

B. Selection of Road Sections

The maintenance data of four roads: Riyadh-Wadi Duwaiser, Riyadh-Taif, Riyadh-Qassim, and Riyadh-Dammam were obtained from the General Directorate of Operations and Maintenance that contain the date of maintenance, maintenance type, and location of maintenance works. Subsequently, pavement sections are selected from these roads. The road sections are selected by considering the following criteria: (1) sections with critical distresses, (2) sections with no overlay and no rehabilitation, and (3) sections with age more than 5 years [11]. Sections with severe distress were defined as having total rutting greater than 6 mm, and IRI value greater than 2.72 m/km. According to the maintenance data, the road sections that have not been maintained are shown in Table II.

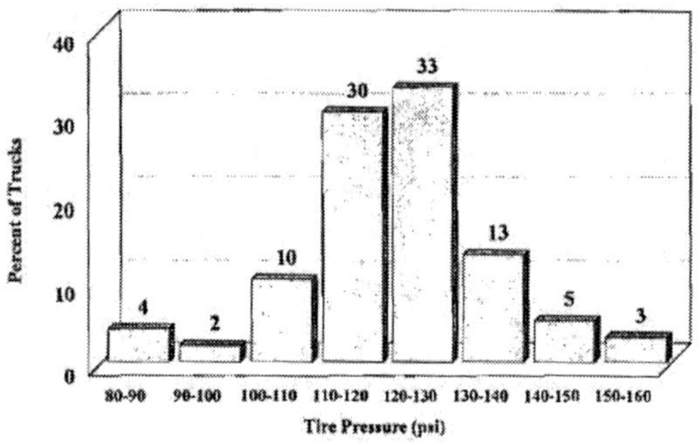

Fig. 1 Distribution of tire pressure values

TABLE II
ROAD SECTIONS THAT HAVE NOT BEEN MAINTAINED

Road Name	From	To	From	To
Riyadh- Wadi Duwaiser	400	572	777	790
	660	663	845	850
	720	730	895	995
	755	757	1010	1039
Riyadh- Taif	730	670	810	765
Riyadh- Dammam	875	840	1020	1015
	980	940		
Riyadh- Qassim	185	180	441	345
	265	225		

C. Pavement Performance

Surveys of the pavement conditions are used to evaluate the pavement performance on the highway network. Automated pavement condition surveys are conducted by using a technologically complex vehicle which has the ability to collect data related to the road's surface distresses, and IRI. The General Directorate of Operations and Maintenance in the Saudi Ministry of Transportation has a project to survey and evaluate the roads which is called the Surveying and Evaluation of the Roads Project. This project uses a complex vehicle (Hawkeye 2000) to measure the road surface distresses using laser technology and digital cameras. The vehicle (Hawkeye 2000) is equipped with a Digital Laser Profiler with 13 laser sensors which can be measured at 10 cm on the road surface, a digital camera on the front part of the vehicle to take images of all right of the road (panorama), a digital camera on the rear of vehicle to take images of the road surface, a device to measure the vehicle's height above the roadway (longitudinal profile) to calculate IRI, and a GPS device. In this paper, the performance data were obtained from the surveying and evaluation of the roads project.

D. Climate Data

Environmental conditions have a significant impact on the performance of flexible pavements because the pavement is exposed to the changes in environmental conditions during the design period. The Enhanced Integrated Climatic Model (EICM) is associated entirely with MEPDG to include the impact of the environmental conditions and seasonal change on pavement performance. The EICM is used to predict the temperature profile through the depth of the pavement with moisture content in the unbound layers. The predicted temperature in flexible pavement correlates directly with the stiffness of the asphalt concrete (AC) layer, and thus, it is important to rutting and thermal cracking. The EICM uses data from climatic files which contain hourly ambient temperature and relative humidity, precipitation, percent of cloud cover, and wind speed [12].

In the Riyadh region, there is a Meteorology and Environment Station at King Khalid International Airport which measures weather conditions. The measurement of weather condition data includes hourly temperature, humidity, wind speed, precipitation, and sky cover (clouds cover), and hence, it is documented in the connecting system with the Presidency of Meteorology and Environment in Jeddah. In this paper, the climate data in the Riyadh region for a period of 25 months (September 1st, 2009 until September 14th, 2011) were obtained from the Presidency of Meteorology and Environment. Groundwater level depths underground vary in the different boreholes in the Riyadh region. In the upper basins, the water level ranges between 40 m and 45 m. From the lower basins, the water level ranges from 30 m to 35 m [13]. Thus, a groundwater depth of 40 m is used for the Riyadh region.

E. Pavement Structure

In the Saudi Ministry of Transportation there was one standard for the structure of flexible pavement of a main road. The typical cross section for the main sections of flexible pavement containing asphalt wearing course with a thickness of 5 cm, asphalt base course with thickness of 20 cm, and a granular base course with a thickness of 12 cm over subgrade [14].

F. Materials Properties

The materials properties were obtained from the Materials and Research Department in the Saudi Ministry of Transportation. This department produced the Saudi Highway Materials Manual which contain the specifications of materials for each layer of a pavement, and will be discussed in accordance with each layer.

1. Asphalt Wearing and Base Course

Based on the Saudi Highway Materials Manual, penetration grade of asphalt binder is 60/70 pen, the percent retained of the sieve analysis for HMA aggregate is given in Table III, and mixture volumetric data are given in Table IV [14].

Reference [15] shows a Poisson's Ratio of 0.35 is used in Riyadh. Typical values of Thermal Conductivity (K) for asphalt concrete range from 0.44 to 0.81 btu/hr-ft-°F. A default value of 0.67 btu/hr-ft-°F is assumed for all of the sections. Typical values of Heat Capacity (Q) for asphalt concrete range from 0.22 to 0.40 btu/lb-°F. A default value of 0.22 btu/ lb-°F is assumed for all of the sections.

TABLE III
HMA AGGREGATE GRADATION

Sieve No.	Asphalt Wearing course	Asphalt Base course
Cumulative % retained 19 mm (3/4") sieve	**0**	**27**
Cumulative % retained 9.5 mm (3/8") sieve	**24**	**47**
Cumulative % retained 4.75 mm (#4) sieve	**48**	**61**
% passing (#200) sieve	**5**	**5**

TABLE IV
MIXTURE VOLUMETRIC DATA

Material Property	Asphalt Wearing course	Asphalt Base course
Voids in the mineral aggregate -VMA (%)	11.85	11.8
Air voids - Va (%)	4.15	3.97
Effective asphalt content (%)	7.7	7.83
Bulk specific gravity - Gmb	2.41	2.36
Total unit weight (density) mix (pcf)	150.46	147.34

2. Granular Base

According to the Saudi Highway Materials Manual, the soil classification of the granular base is "A-1-a" based on the AASHTO Soil Classification System. The value of modulus is 48,700 psi. The Plastic Index is six as a maximum.

3. Subgrade

Al-Suhaibani et al. [16] investigated subgrade soil of all roads in Saudi Arabia. They found properties of subgrade soil in the Riyadh region: soil classification is "A-2-4" based on the AASHTO Soil Classification System. The value of modulus is 14,200 psi [16].

IV. DATA ANALYSIS

The collected data need to be developed to be suitable as input data for MEPDG. The development of the input data parameters is described in the next paragraphs.

A. Traffic Characteristics

1. Vehicle Classification

The MEPDG uses as default the Federal Highway Administration (FHWA) standard classification scheme that classifies vehicles into 13 different classes. Based on this classification, vehicle class 1 is motorcycles, vehicle class 2 is passenger cars, vehicle class 3 is pickup trucks, vehicle class 4 is buses, and vehicle classes 5 through 13 are trucks [17]. The Saudi vehicle classification standard in traffic counting stations classifies vehicles into 15 classes. According to this classification, class 1 is passenger cars and pickups, classes 2 through 14 are trucks, and class 15 is unclassified vehicles.

The Saudi trucks classification is unlike the default truck classification in MEPDG. Therefore, the adaption of Saudi truck classification to the MEPDG default is needed. The differences between the Saudi trucks classification and the MEPDG default classification were minimized based on truck type, the number of axles, and axle distances (single axle, tandem axle, tridem axle). Subsequently, Saudi trucks classification was converted as MEPDG default.

Also, the Saudi truck classification in truck weigh stations was converted to the default truck classification in MEPDG, as it is different from both the Saudi trucks classification in traffic counting stations and the MEPDG default classification. Table V shows the Saudi trucks classification converting to the MEPDG default.

TABLE V
CONVERTING OF SAUDI TRUCKS TO FHWA TRUCKS CLASSIFICATION

MEPDG trucks classification	Saudi trucks classification	Saudi trucks weight stations classification	No. Axles
4	2	--	3
5	3	1	2
6	5	2	3
7	6+10	9+10	4
8	4+7	3+4+6	3+4
9	9	7	5
10	11+12+14	8	6
11	8	5	5 (Trail+3)
12	–	–	6 (Trail+4)
13	13	99	7 (Trail+5)

The last Saudi vehicles classification is unclassified vehicles, where the traffic count device failed to identify the vehicle types based on the integrated criteria. They include only the number of vehicles without any other measurements such as axle number and axle distances. The possible reason for this is that it includes irregular trucks. Unclassified vehicles should not be neglected as they affect the pavement design, and therefore, their volumes were added to the total truck volumes [18].

TABLE VI
TRAFFIC GROWTH FACTOR

Year	AADT	Growth Factor
2011	82847	
2012	92850	1.121
2013	96983	1.045
2014	99681	1.028
2015	104298	1.046
	AVR.	1.060

2. Traffic Growth Factors

The annual average daily traffic (AADT) for the period 2011 to 2015 were obtained from the permanent traffic counting stations and average growth factors were developed as shown in Table VI.

3. Annual Average Daily Truck Traffic

It is the average volume of truck traffic (heavy vehicles classes 4 to 13 in the traffic stream) passing a point or segment of a road in both directions over a period of 24-hours. The MEPDG allows for traffic input to be either entered directly as Average Annual Daily Truck Traffic (AADTT) or can be calculated using the AADTT calculator, where the Average

Annual Daily Traffic (AADT) and percent of trucks are multiplied [19].

The Average Annual Daily Traffic (AADT) in the year 2012 for each road was obtained from the General Directorate of Operations and Maintenance in the Saudi Ministry of Transportation. Thereafter, it was converted to the AADT of the base year by using (1), as shown in Table VII.

$$AADTT_{BY} = AADTT_X \div (GR)^n \quad (1)$$

where $AADTT_{BY}$ *is* annual average daily truck traffic at age X, $AADTT_X$ is base year annual average daily truck traffic, *n* is the number of years from the base year (date of construction) to the year 2011, and *GR* is the traffic growth rate of 1.06.

TABLE VII
CONVERSION OF AADT 2012 TO THE AADT OF BASE YEAR

No.	Road Name	AADT (2012) (veh/d)	Date of const.	AADT of base year (veh/d)	Trucks (%)	AADTT (veh/d)
10	Riyadh-Wadi Duwaiser	20745	1981	3407	29	988
40	Riyadh-Taif	30510	1981	5011	31	1553
65	Riyadh-Qassim	22245	1982	3873	22	852
40	Riyadh-Dammam	19350	1981	3178	44	1398

4. Percent of Trucks in Design Direction

The percent of trucks is commonly assumed to be 50 percent when the AADT and AADTT are given in two directions as it is used [9].

5. Percent of Trucks in the Design Lane (LDF)

The design lane is typically the outside lane of a multilane highway. The default values recommended for use are [9]:

- Single-lane roadways in one direction, LDF = 1.00
- Two-lane roadways in one direction, LDF = 0.90
- Three-lane roadways in one direction, LDF = 0.60
- Four-lane roadways in one direction, LDF = 0.45

The chosen roads in Riyadh region have three lanes in each direction, so the lane distribution factor (LDF) is 0.60.

6. Vehicle Operational Speed

Reference [20] shows the classification of operational speed on highway type. So, the vehicle operational speed of 100 km/hr (60 mph) is selected for the Riyadh region.

7. Monthly Adjustment Factors

It is defined as the monthly variation in truck traffic throughout the entire year. Equation (2) was used to calculate these factors for Riyadh region which can be shown in Fig. 2.

$$MAF_i = \frac{AMDTT_i}{\sum_{i=1}^{12} AMDTT_i} \times 12 \quad (2)$$

where *MAFi* is monthly adjustment factors for month*i*, and *AMDTTi* is average monthly daily truck traffic for month*i*.

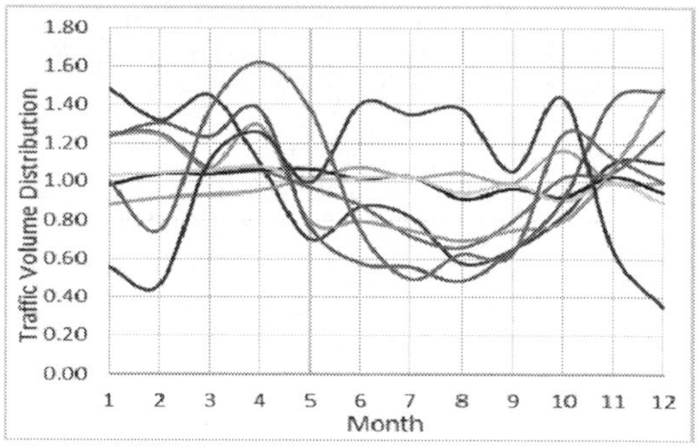

Fig. 2 Monthly Adjustment Factors Variation

Fig. 2 shows that the fewer trucks travel during periods of poor driving conditions in the summer season due to high temperatures, and most trucks travel during good driving conditions.

8. Vehicle Class Distribution

Vehicle class distribution factors represent the percent of each truck class type within the AADTT. The vehicle class distribution for the Riyadh region was calculated by (3).

$$VCD_i = \frac{AADTT_i}{AADTT} \times 100 \quad (3)$$

where VCD_i is the vehicle class distribution factor for vehicle class*i*, $AADTT_i$ is the annual average daily truck traffic for class*i*, and $AADTT$ is annual average daily truck traffic for all classes.

Practically, when the calculation has been completed for all the vehicles classes, the sum of all the vehicle class distributions should be equal 100% [21]. If this is not achieved, there may be an error in data collection or analysis.

Fig. 3 depicts the vehicle class distribution which is selected for use in the MEPDG in the Riyadh region.

As can be seen, vehicle class 5, 7, and 8 were noted to be the most common in the Riyadh region highway system.

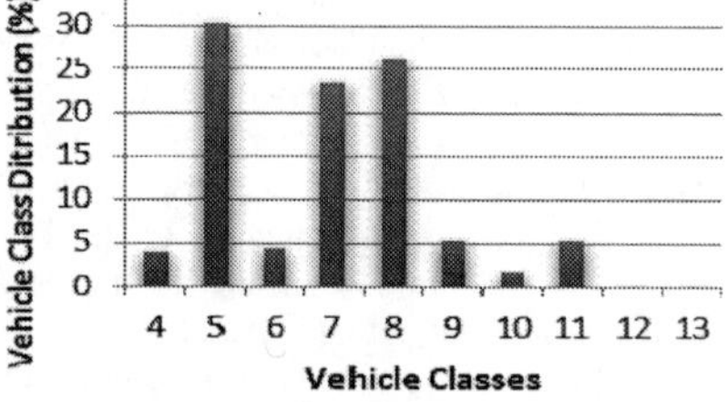

Fig. 3 Riyadh Region Vehicle Class Distribution

9. Truck Hourly Distribution Factors

Equation (4) was used to compute the hourly distribution factors for Riyadh region.

$$HDF_i = \frac{AAHT_i}{\sum_{i=1}^{24} AAHT_i} \times 100 \quad (4)$$

where *HDFi* is hourly distribution factor for *i*th one-hour time period, and *AAHTi* is annual average hourly traffic for *i*th one-hour time period.

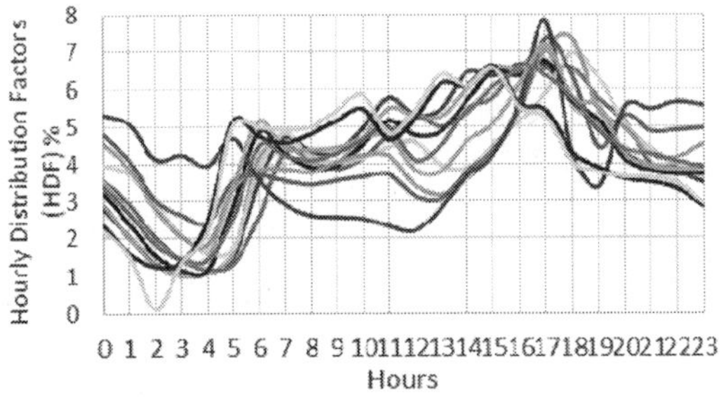

Fig. 4 The Variations of the Hourly Distribution Factors

The *HDFi* are input into the MEPDG as a percent. Therefore, the summation of the 24-hourly distribution factors must be equal to 100.

Fig. 4 shows the variations of the hourly distribution factors at the site and indicates the highest value of these factors at the hour of 17 for most months.

10. Axle Load Distribution Factors

The axle load distribution factors (axle load spectra) represent the percent of the axles that are sorted with each load increments for each axle type (single, tandem, tridem, and quad) and vehicle class (classes 4 through 13).

The load ranges and intervals for each axle type are:

- Single axles – 1361 kg to 18597 kg at 454 kg intervals.
- Tandem axles – 2722 kg to 37195 kg at 907 kg intervals.
- Tridem and quad axles – 5443 kg to 46266 kg at 1360 kg.

Equation (5) was used to calculate the axle load distribution for the Riyadh region.

$$ALDF_{ijk} = \frac{\text{No. of axles}_{ijk}}{\text{Total No. of axles}_{ij}} \times 100 \quad (5)$$

where *ALDF* is axle load distribution factor, *i* represents the vehicle class (4 through 13), *j* the month (1 through 12) and *k* the load range (1361 kg to 18597 kg at 454 kg intervals for single axles). This is done for each axle type (single, tandem, tridem, and quad). These procedures were performed for each month (January to December), each class of trucks (4 - 13), and each type of axles (single, tandem and tridem). Fig. 5 shows the examples of the axle load distributions calculated for the month of January. In these procures, the vehicles with quad axles were not found within the obtained data from trucks weight stations, so the quad axle load distribution factors were set equal to zero. In addition, the buses (trucks class 4) were not measured in these stations, and therefore, the default axle load distribution factors are used for this class. Furthermore, the truck class 12 does not exist in the Saudi trucks classification, thus the axle load distribution factors of this class were set to zero.

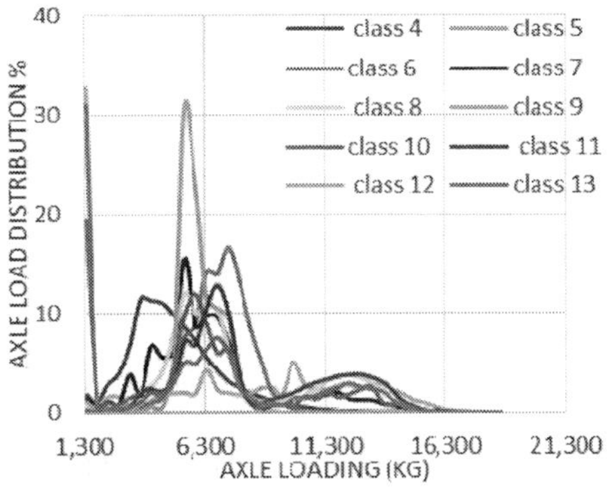

Fig. 5 Single Axle Load Distributions for January month

11. General Traffic Inputs

- Mean Wheel Location: This is the distance from the outer edge of the wheel to the lane longitudinal pavement marking. The MEPDG default value of 460 mm is used for the Riyadh region inputs.
- Design Lane Width: The design lane width is the actual traffic lane width defined by the distance between the pavement markings on both sides of the design lane. In the Riyadh region, the design lane width is 3.65 m.
- Number of Axle Types per Truck Class: The number of axle types per truck class is the average number of individual axles for each truck class for axle type (single, tandem, tridem, and quad). This number was calculated by dividing the sum of each axle type for each truck class by sum of the trucks in that class. This procedure was performed for single, tandem, and tridem axles for each truck class (5-13) and each month (January to December) as can be seen in Table VIII. The number of axles for class 4 was not measured, so the default value of MEPDG is used.

TABLE VIII
THE NUMBER OF AXLE TYPES PER TRUCK CLASS FOR RIYADH REGION

Truck Class	Number of Single Axles per Truck	Number of Tandem Axles per Truck	Number of Tridem Axles per Truck	Number of Quad Axles per Truck
4	1.62	0.39	0.00	0.00
5	2.00	0.00	0.00	0.00
6	1.00	1.00	0.00	0.00
7	1.76	0.27	0.56	0.00
8	2.00	0.98	0.00	0.00
9	1.20	1.19	0.03	0.00
10	2.25	0.50	0.77	0.00
11	2.01	0.17	0.99	0.00
12	0.00	0.00	0.00	0.00
13	3.99	0.58	0.53	0.00

- Axle Configuration: This input includes several variables:
  - ❖ Average axle width (m): It is the distance between two outside edges of an axle. MEPDG default value of 2.60 m is used as the average axle width.
  - ❖ Dual tire spacing (mm): It is the distance between the centers of a dual tire. MEPDG default value of 305 mm is used in the pavement sections.
  - ❖ Tandem axle spacing (mm): It is the center-to-center longitudinal spacing between two consecutive axles in a Tandem configuration. The local value of 2.5 m is used in the Riyadh region inputs.
  - ❖ Tridem axle spacing (mm): It is the center-to-center longitudinal spacing between two consecutive axles in a tridem configuration. The local value of 2.5 m is used in the Riyadh region inputs.

These variables are important in describing the loads applied to the pavement because pavement structure is sensitive to the distance between wheels and axles within a truck.

- Wheelbase: The wheelbase refers to the distance between the steering axle and the first axle of the truck. It has been characterized as short, medium, and long axle spacing.

The average spacing of short axles (m): This defines the average longitudinal spacing of short axles. MEPDG default value of 3.66 m is used.

The average spacing of medium axles (m): This defines the average longitudinal spacing of short axles. MEPDG default value of 4.6 m is used.

The average spacing of long axles (m): This defines the average longitudinal spacing of short axles. MEPDG default value of 5.49 m is used.

The details of the wheelbase of the truck are needed for use in computing pavement responses.

Typically, short trucks are Class 5, medium single unit trucks are Class 6 and Class 7, and long trucks are Classes 8 to 13. The Percent of trucks for wheelbase as can be seen in Table IX.

TABLE IX
PERCENT OF TRUCKS FOR WHEELBASE

Type	Percent of trucks
Short	32.75 %
Medium	35.74 %
Long	31.51 %

B. Climate Information

The data for hourly temperature, wind speed, precipitation and relative humidity can be directly used from the obtained data. The data for cloud cover needs to be transposed. The MEPDG climate input requires a percent of cloud cover, but the Presidency of Meteorology and Environment in Riyadh uses the okta unit to define cloud cover. An okta is a measurement unit which is used to represent the quantity of cloud cover. In this scale, the sky is divided into eights parts and the cloud cover are estimated in term of how many oktas (eights) the cloud cover the sky, ranging from 0 (completely clear sky) through to 8 oktas (completely overcast), as shown in Fig. 6 [22].

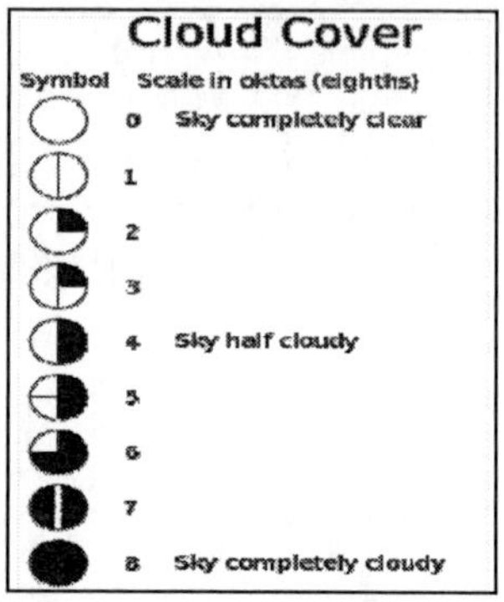

Fig. 6 Scale of cloud cover measured in Oktas (eighths)

In this paper, the one okta was considered 12.5% to convert okta unit into percentage to define cloud cover. This means, 1 okta is 12.5%, 2 okta are 25%, 3 okta are 37.5%, 4 okta are 50%, 5 okta are 62.5%, 6 okta are 75%, 7 okta are 87.5%, and 8 okta are 100%. A simple excel sheet was designed for Riyadh climate data and it was converted to a text-file with an extension ".hcd" as required in the MEPDG.

V. CONCLUSIONS

The conclusions from this paper are:

1) The Saudi trucks classification is unlike the default truck classification in MEPDG, and therefore, the Saudi truck classification was modified to the MEPDG default based on the number of axles and axle spacing.
2) Truck class 12 in MEPDG default was not used because it does not exist in the Saudi truck classification.
3) Unclassified vehicles should not be neglected as they effect the pavement design, and so, he unclassified vehicle volumes were added to the total truck volumes.
4) The monthly adjustment factors indicate that the truck traffic increases in the winter season and conversely in the summer season, this means that fewer trucks travel during periods of poor driving conditions of the summer season due to the high temperature, and most of them travel during good driving condition.
5) Vehicle class distribution indicates that the vehicle classifications 5, 7, and 8 are the most common in the Riyadh region highway system.
6) Truck Hourly Distribution denotes that most trucks travel at 5:00 pm for most months.
7) Through the processing of axle load distribution factors (axle load spectra), the trucks with quad axles were not found within the obtained data from truck weight stations, so the quad axle load distribution factors are set equal to zero.
8) The buses (trucks class 4) were not measured in the truck weigh stations and therefore, the default axle load distribution factors are used for this class.

REFERENCES

[1] C. W. Schwartz and R. Carvalho, "Evaluation of mechanistic-empirical design procedure," *Final Report, MDSHA Project No. SP0077B41, Maryland,* 2007.

[2] K. Wang, Q. Li, K. Hall, V. Nguyen, W. Gong, and Z. Hou, "Database Support for the New Mechanistic-Empirical Pavement Design Guide," *Transportation Research Record: Journal of the Transportation Research Board,* pp. 109-119, 2008.

[3] R. Delgadillo, C. Wahr, and J. Alarcón, "Toward Implementation of the Mechanistic-Empirical Pavement Design Guide in Latin America: Preliminary Work in Chile," *Transportation Research Record: Journal of the Transportation Research Board,* pp. 142-148, 2011.

[4] A. Ghosh, A. Padmarekha, and J. M. Krishnan, "Implementation and proof-checking of mechanistic-empirical pavement design for Indian highways using AASHTOWARE pavement ME design software," *Procedia-Social and Behavioral Sciences,* vol. 104, pp. 119-128, 2013.

[5] Y. Zhao, Y. Tan, and C. Zhou, "Determination of axle load spectra based on percentage of overloaded trucks for mechanistic-empirical pavement design," *Road Materials and Pavement Design,* vol. 13, pp. 850-863, 2012.

[6] L. G. Loria, G. Badilla, M. Jimenez Acuna, F. Elizondo, and J. P. Aguiar-Moya, "Experiences in the Characterization of Materials Used in the Calibration of the AASHTO'Mechanistic-Empirical Pavement Design Guide (MEPDG) for Flexible Pavement for Costa Rica," in *Transportation Research Board 90th Annual Meeting,* 2011.

[7] C. Caliendo, "Local calibration and implementation of the mechanistic-empirical pavement design guide for flexible pavement design," *Journal of Transportation Engineering,* vol. 138, pp. 348-360, 2011.

[8] H. Sadek, E. Masad, O. Sirin, H. Al-Khalid, and D. Little, "The implementation of mechanistic-empirical pavement design method to evaluate asphalt pavement design in Qatar," in *5th Eurasphalt & Eurobitume Congress,* 2012, pp. 13-15.

[9] M.-E. P. D. Guide, "Guide for mechanistic empirical design of new and rehabilitated pavement structures," *NCHRP Rep. 1-37A, Final Rep,* 2004.

[10] A. I. Al-Mansour and E. A. Sharaf, "Analysis of Current Truck Tire Pressure Levels and Their Effects on Highway Pavements in Saudi Arabia," 1995.

[11] M. Kang and T. M. Adams, "Local calibration for fatigue cracking models used in the Mechanistic-empirical pavement design guide," in *Proceedings of the 2007 Mid-Continent Transportation Research Symposium,* 2007.

[12] J. Saha, S. Nassiri, A. Bayat, and H. Soleymani, "Evaluation of the effects of Canadian climate conditions on the MEPDG predictions for flexible pavement performance," *International Journal of Pavement Engineering,* vol. 15, pp. 392-401, 2014.

[13] M. S. Kalthem, "Evaluation of Riyadh City water supply and demand," 1978.

[14] H. I. Al-Abdul Wahhab, M. Fatani, A. Noureldin, A. Bubshait, and I. AL-DUBABE, "National Study of Asphalt Pavement Rutting in Saudi Arabia," *Transportation Research Record,* 1995.

[15] M. H. Alawi, & Backar A. H., "Thermo-Mechanical model of flexible pavement structure behavior," 2010.

[16] A. Al-Suhaibani, T. Al-Refeai, and A. Noureldin, "Characterization of Subgrade Soil in Saudi Arabia; A study of Resilient Behavior," *KACST Project No. AR-12-51, Final report,* 1997.

[17] A. R. Abbas and A. Frankhouser, "Improved characterization of truck traffic volumes and axle loads for mechanistic-empirical pavement design," 2012.

[18] Y. Jiang, S. Li, T. Nantung, K. Mangold, and S. A. MacArthur, "Creation of Truck Axle Load Spectra Using Weigh-in-Motion Data," in Journal of the transportation Research Forum, 2010.

[19] P. Interactive, "Roughness," Pavement Interactive http:// www.pavementinteractive.org/article/roughness. Accessed on 12/11/2016.

[20] A. Aashto, "Policy on geometric design of highways and streets," *American Association of State Highway and Transportation Officials, Washington, DC,* vol. 1, p. 158, 2001.

[21] T. Kasperick and K. Ksaibati, "Calibration of the Traffic Distributions and Model Coefficients within the MEPDG for Local Energy-Affected Roads in Wyoming," in *Transportation Research Board 92nd Annual Meeting,* 2013.

[22] Chaston, P.R. "Weather Maps: How to Read and Interpret All the Basic Weather Charts," Chaston Scientific Incorporated, 2009.

Preparation of n-type Bi_2Te_3 Films by Electrophoretic Deposition

Tahereh Talebi, Reza Ghomashchi, Pejman Talemi, Sima Aminorroaya

***Abstract*—**A high quality crack-free film of Bi_2Te_3 material has been deposited for the first time using electrophoretic deposition (EPD) and microstructures of various films have been investigated. One of the most important thermoelectric (TE) applications is Bi_2Te_3 to manufacture TE generators (TEG) which can convert waste heat into electricity targeting the global warming issue. However, the high cost of the manufacturing process of TEGs keeps them expensive and out of reach for commercialization. Therefore, utilizing EPD as a simple and cost-effective method will open new opportunities for TEG's commercialization. This method has been recently used for advanced materials such as microelectronics and has attracted a lot of attention from both scientists and industry. In this study, the effect of media of suspensions has been investigated on the quality of the deposited films as well as their microstructure. In summary, finding an appropriate suspension is a critical step for a successful EPD process and has an important effect on both the film's quality and its future properties.

***Keywords*—**Bi_2Te_3, electrical conductivity, electrophoretic deposition, thermoelectric materials, thick films.

I. Introduction

GLOBAL energy consumption is increasing rapidly. To ensure a sustainable future, there is a need to develop better ways to produce and utilise energy. Almost all of the energy production methods, as well as industrial processes, generate heat that cannot be used and will be wasted and released into the atmosphere, increasing the threat of global warming. Nearly 60% of the energy produced by mankind is in the form of wasted thermal energy [1]. If this enormous amount of energy could be harnessed into more useful energy forms, it would reduce greenhouse gases and fossil fuel consumption, significantly reducing the risks of global warming [2], [3].

The solution for this problem may lie in using TE materials, which are capable of converting waste heat into electricity. The waste heat in systems such as power plants, cars, planes, computers or even stovetops can be sources of electricity. TE materials could also be used for refrigeration purposes, without the shortcomings associated with vibration in conventional chillers [4], [5].

When a thermal gradient is applied to TE materials, it will generate electricity. The efficiency of this conversion relies on the magnitude of the temperature difference, and intrinsic properties of the TE materials, as well as the design of the TE generator in terms of its shape, electric contact, etc. [2].

Before the commercialization of these devices, various challenges have to be solved by researches and industry to increase their uses from limited niche applications, such as deep-space satellites, high temperature waste heat recovery and remote power generation for unmanned systems [1], [6]. Two of the most important challenges are the high weight of conventional TE devices and the high cost of manufacturing them [1], [7], [8].

To lower the weight of these devices, especially for mobile applications, fabricating TE films has been the focus of current research in the industry [9]. TE films can have interesting and varied applications. For instance, body heat can be harnessed to recharge a mobile phone, iPad or other personal gadgets [10]. Therefore, the feasibility of TE devices heavily depends on the cost and energy conversion efficiency. Taking new approaches toward TE materials using nanostructures and novel crystal structures has resulted in a rapid development of TE materials with significant higher energy efficiencies [1], [5], [12].

Thermoelectric materials have various classifications such as Half-Heusler materials, Clathrates, skutterudites, Chalcogenides which have different characteristics [2], [7], [16].

Bismuth Telluride, the most studied TE material, is a simple TE material which is commercially available and has interesting TE properties.

Fig. 1 shows the Bi-Te binary phase diagram [4]. The arrow shows the position of Bi_2Te_3 in the diagram. Its Seebeck coefficient depends on its composition (Te deficiency) and doping materials. The maximum ZT values reported for pure crystalline Bi_2Te_3 materials are about 0.75 and 0.86 for p-type and n-type, respectively [11].

As an advantage, a complete solid solubility exists among Bi_2Te_3, Sb_2Te_3 and Bi_2Se_3 which all have noble TE features. Therefore, ZT and other TE properties can be improved by the addition of Sb_2Te_3 and Bi_2Se_3 to Bi_2Te_3 [3]. Bi_2Te_3 can form p-type and n-type semiconductor solid solutions simply by adding Sb_2Te_3 or Bi_2Se_3, resulting in $Bi_{2-x}Sb_xTe_3$ and $Bi_2Te_{3-x}Se_x$ alloys, respectively [5]. This will result in lowering thermal conductivity but not degrading electrical conductivity. The optimum compositions with ZT~1 at room temperature

T. Talebi is with the Mechanical Engineering School, the University of Adelaide, South Australia, Australia (phone: 883-133-360; fax: 883-134-367; e-mail: tahereh.talebinamakrodbari@adelaide.edu.au).

R. Ghomashchi is with the Mechanical Engineering School, the University of Adelaide, South Australia, Australia (e-mail: reza.ghomashchi@adelaide.edu.au).

P. Talemi is with the Chemical Engineering School, the University of Adelaide, South Australia, Australia (e-mail: pejman.talemi@adelaide.edu.au).

S. Aminorroaya is with Institute for Superconducting & Electronic Materials, University of Wollongong, New South Wales, Australia (e-mail: sima@uow.edu.au).

are usually $Bi_2Te_{2.7}Se_{0.3}$ (n-type) and $Bi_{0.5}Sb_{1.5}Te_3$ (p-type) [4], [8].

To decrease the cost of production, a cost-effective film fabrication method, called EPD has been used. This method is a simple non-vacuum technique with a high rate of deposition method to prepare thick films [11], [12].

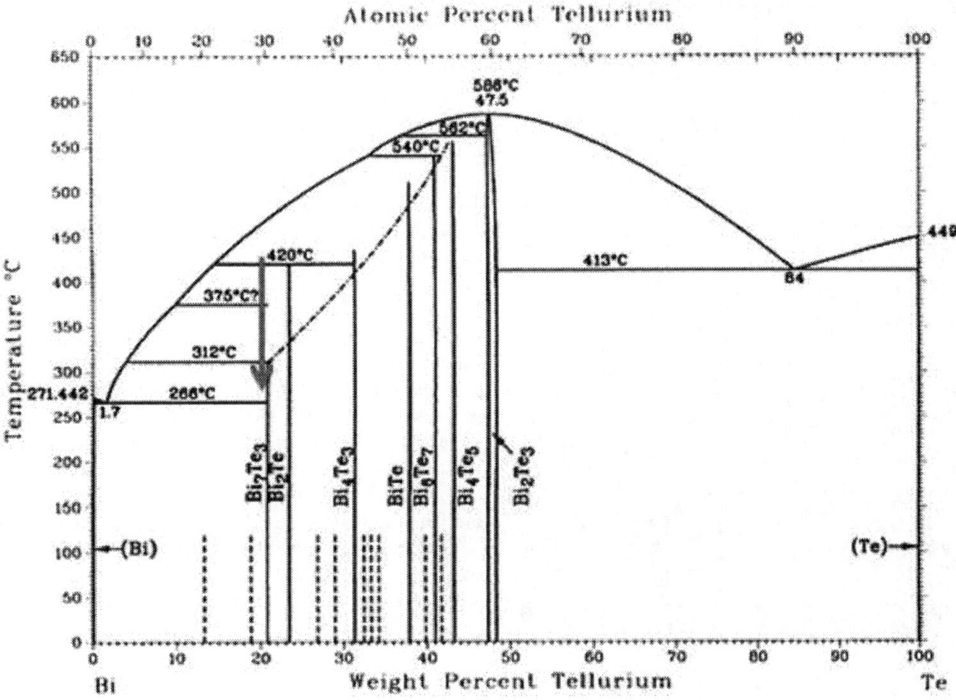

Fig. 1 The Bi-Te binary phase diagram [4]

In EPD, a DC current electric field is applied to a suspension of the charged powder particles, and then they are attracted and deposited on the substrate with the opposite charge. Different factors from both the suspension and EPD process can influence the deposited film [11], [13]. EPD has been used with numerous materials for a wide range of applications such as solid oxide fuel cells [14], functionally graded ceramics [15], superconductors [16] etc.

Determining the appropriate media for a suspension is a critical step to achieve a homogenous and dense film in EPD [17]. Therefore, in this research, various suspensions with different media have been prepared and the quality, the macrostructure and the microstructure of the EPD films have been investigated.

II. Experimental Procedure

Bi_2Te_3 powder was commercially available from the Bio-chemical Leader Company with 99.99% purity. The chemical composition of p-type Bi_2Te_3 powder was 22.51 Bi (wt %), 54.90 Te (wt %) and 23.59 Sb (wt %).

The median particle size (D50) of 20μm was characterized for particle sizes of the powder using a Malvern Mastersizer 2000 laser particle size analyser. Scanning electron microscope (SEM - Quanta 450) has been utilized to obtain the SEM images of the starting powder is shown in Fig. 2 at diverse magnifications.

To prepare the suspension, 1g of Bi_2Te_3 powders were dispersed in 100 ml of the media mixture (10 g/L Bi_2Te_3 suspension). The mixture was put in an ultrasonic bath for 15 min to disperse and break the agglomerates and stabilise the suspension. The suspension was then rested for at least for 2 hours to reach stability. The suspension was ultrasonicated for 2 minutes prior to the EPD process.

All of the media supplied by the Sigma Aldrich Company and the purity of the ethanol and ethylene glycol was 99.8%; the purity of the acetone and isopropanol was 99.5%; and the purity of the tetrahydrofuran and acetylacetone was 99.9%, 99%, respectively.

A schematic image of the EPD cell is shown in Fig. 3. The distance between the substrate and the counter electrode was constant at 10 mm for the EPD process. The electrodes were vertical and kept parallel and fixed during the EPD process.

The substrate and the counter electrode were cut from a commercially obtained copper foil roll, with dimensions of 10 mm by 10 mm.

Fig. 2 The SEM image of p-type Bi_2Te_3 powder at diverse magnifications (1000x and 5000x)

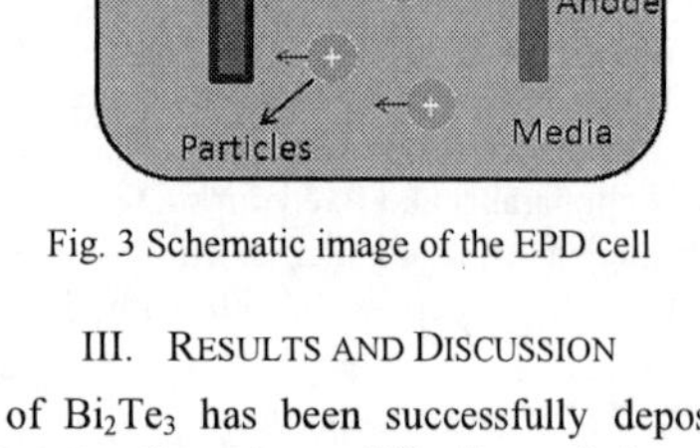

Fig. 3 Schematic image of the EPD cell

III. Results and Discussion

A film of Bi_2Te_3 has been successfully deposited on a copper substrate through modification of the suspension mixture. To achieve a high quality, uniform film, more adjustments and experiments were needed to reach a fundamental understanding of the effect of different parameters on the deposited films and their characteristics, as will be discussed in the following sections.

To reach the highest efficiency and figure of merit (ZT) for a TE material, the total thermal conductivity should be as low as possible. However, the thermal conductivity of metals is proportional to its electrical conductivity because it is dominated by electrons. The electrical conductivity of TE materials directly depends on its electronic thermal conductivity (k_e) which is a function of its band structure [1]. Thus, the lattice thermal conductivity (k_L) is the only parameter that can be reduced without too much negative effect on the electrical conductivity and figure of merit of TE materials. It is believed that phonons are responsible for transferring lattice thermal conductivity [3].

Thermal conductivity (k) consists of two of electronic (k_e) and lattice (k_L) components and T is the absolute temperature [6]. The power factor also has a significant effect on the performance of TE materials. The figure of merit can be increased by enhancing power factor and decreasing thermal conductivity. Semiconductors have the highest power factor amongst the materials because electrical conductivity of insulators and Seebeck coefficient of metals are too low [2].

Therefore, the mobility of phonons must be minimised whereas the mobility of electrons have to remain as high as possible to achieve the highest ZT value .It also has been suggested that the highest figure of merit will be accomplished in a material with the electrical properties of a crystal while it has the thermal properties of a glass which is known as the Phonon-Glass Electron-Crystal (PGEC) material [5].

A. The Effect of the Suspension on the Weight of Deposition

Fig 4 shows the deposition weights per area of suspensions with medium and high quality depositions at 100V for 9 minutes. The results show that the suspension with a higher zeta potential and electrophoretic mobility yielded a higher deposition weight, which is in good agreement with EPD theoretical principles.

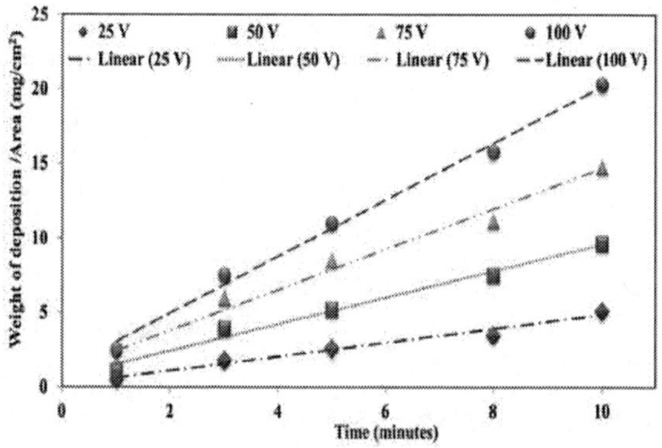

Fig. 4 The deposition weights per area of different suspensions with medium or high quality depositions.

As discussed before, according to Hamaker's equation [11]-[13], the zeta potential and electrophoretic mobility have a direct dependence on the deposition weights. These relationships have also been confirmed by the experimental results, which show a similar trend for the deposition weights to the ones of zeta potential and electrophoretic mobility.

B. The Effect of EPD Time on the Film Microstructure

Fig. 5 I shows the SEM images of two green Bi_2Te_3 films deposited from a THF suspension at 100V/cm for (a) 5 minutes and (b) 10 minutes. It illustrates that when the deposition time doubles; the microstructure of the film also almost differs, which is in line with the measured results from the weight of the deposition experiments and Hamaker's equation as well as the EPD theoretical principle.

(a)

(b)

Fig. 5 The SEM images of green Bi_2Te_3 films at two different EPD voltages, (a) 100V and (b) 50V for 10 minutes from the 75% vol. Ethanol + 25% vol. Acetone suspensions

IV. Conclusion

In summary, a film of Bi_2Te_3 has been successfully deposited on a copper substrate and the effects of EPD time on the film microstructure as well as the effect of the suspension on the weight of deposition have been investigated. The result shows that time of the process and type of the suspension have significant effect on the deposited films.

Acknowledgment

This work has been supported by the Australian Government Research Training Program Scholarship and The University of Adelaide.

References

[1] DiSalvo FJ. Thermoelectric cooling and power generation. Science. 1999;285:703-6.
[2] Snyder GJ, Toberer ES. Complex thermoelectric materials. Nature Materials. 2008;7:105-14.
[3] Shakouri A. Recent developments in semiconductor thermoelectric physics and materials. Annual Review of Materials Research. 2011;41:399-431.
[4] Dresselhaus MS, Chen G, Tang MY, Yang RG, Lee H, Wang DZ, et al. New directions for low-dimensional thermoelectric materials. Advanced Materials. 2007;19:1043-53.
[5] Hamid Elsheikh M, Shnawah DA, Sabri MFM, Said SBM, Haji Hassan M, Ali Bashir MB, et al. A review on thermoelectric renewable energy: Principle parameters that affect their performance. Renewable and Sustainable Energy Reviews. 2014;30:337-55.
[6] LeBlanc S, Yee SK, Scullin ML, Dames C, Goodson KE. Material and manufacturing cost considerations for thermoelectrics. Renewable and Sustainable Energy Reviews. 2014;32:313-27.
[7] Alam H, Ramakrishna S. A review on the enhancement of figure of merit from bulk to nano-thermoelectric materials. Nano Energy. 2013;2:190-212.
[8] Chen S, Ren Z. Recent progress of half-Heusler for moderate temperature thermoelectric applications. Materials Today. 2013;16:387-95.
[9] Nolas GS, Poon J, Kanatzidis M. Recent developments in bulk thermoelectric materials. MRS Bulletin. 2006;31:199-205.
[10] Yang J, Caillat T. Thermoelectric Materials for Space and Automotive Power Generation. MRS Bulletin. 2006;31:224-9.
[11] Besra L, Liu M. A review on fundamentals and applications of electrophoretic deposition (EPD). Progress in Materials Science. 2007;52:1-61.
[12] Corni I, Ryan MP, Boccaccini AR. Electrophoretic deposition: From traditional ceramics to nanotechnology. Journal of the European Ceramic Society. 2008;28:1353-67.
[13] Hamaker HC. The London—van der Waals attraction between spherical particles. Physica. 1937;4:1058-72.
[14] Talebi T, Raissi B, Maghsoudipour A. Electrophoretic deposition of YSZ electrolyte on porous NiO-YSZ substrate for solid oxide fuel cells. Key Engineering Materials. 2009;412:215- 20.
[15] Put S, Vleugels J, Anné G, Van der Biest O. Functionally graded ceramic and ceramic–metal composites shaped by electrophoretic deposition. Colloids and Surfaces A: Physicochemical and Engineering Aspects. 2003;222:223-32.
[16] Maiti HS, Datta S, Basu RN. High-Tc Superconductor Coating on Metal Substrates by an Electrophoretic Technique. Journal of the American Ceramic Society. 1989;72:1733-5.
[17] Javidi M, Bahrololoom ME, Javadpour S, Ma J. Studying surface charge and suspension stability of hydroxyapatite powder in isopropyl alcohol to prepare stable suspension for electrophoretic deposition. Advances in Applied Ceramics. 2009;108:241-8.

Computer Aided Assembly Attributes Retrieval Methods for Automated Assembly Sequence Generation

M. V. A. Raju Bahubalendruni, Bibhuti Bhusan Biswal, B. B. V. L. Deepak

***Abstract*—**Achieving an appropriate assembly sequence needs deep verification for its physical feasibility. For this purpose, industrial engineers use several assembly predicates; namely, liaison, geometric feasibility, stability and mechanical feasibility. However, testing an assembly sequence for these predicates requires huge assembly information. Extracting such assembly information from an assembled product is a time consuming and highly skillful task with complex reasoning methods. In this paper, computer aided methods are proposed to extract all the necessary assembly information from computer aided design (CAD) environment in order to perform the assembly sequence planning efficiently. These methods use preliminary capabilities of three-dimensional solid modelling and assembly modelling methods used in CAD software considering equilibrium laws of physical bodies.

***Keywords*—**Assembly automation, assembly attributes, assembly sequence generation, computer aided design.

I. Introduction

OVER the past four decades, industrial engineers are striving to solve assembly sequence planning problem efficiently with the continuous technology advancements in order to reduce the overall assembly cost. In early 90s, assembly sequence planning problem solving methods were developed aiming to generate single or numerous valid assembly sequences [1]-[3]. These methods are highly time consuming and need extremely skilled user's intervention to answer assembly relational queries [4]-[6]. Bourjault [1] and De Fazio's liaison based precedence relation method [3] and Homem de Mello's assembly cut-set method [2] belong to this category.

Owing to possibility of several feasible assembly sequences for a product, researcher's attention is moved towards achieving optimal assembly sequence in less computational time. Researchers used various artificial intelligence (AI) techniques like Genetic algorithms, Simulated annealing, Neural networks, Particle Swarm Optimization, Harmonious colony-decision algorithm, Ant colony algorithm, Hybrid algorithms, and many others to achieve optimal solution efficiently [7]-[11]. Assembly product information is the primary parameter for these AI based assembly sequence planning problems.

Each assembly sequence generated during assembly sequence planning must be tested for all necessary assembly predicates. However, the manual mode of extracting the assembly information is highly time consuming and skillful task for any product with large number of parts. Many studies have been done to extract these data from CAD based products using basic features and application programming interface (API) compatibility [12], [13]. Liaison matrix and assembly connection matrix extraction from CAD models were discussed by researchers using assembly contact information [14]-[16]. Geometrical feasibility test can be done by using part mating features and assembly interference matrices. Extraction of part mating features from different CAD data exchange formats were reported [17]-[19]. Collision detection algorithms are best used to generate the assembly interference matrices [20]. Gu and Yan [21] and Mok et al. [22] made attempts to generate assembly interference matrices from different formats of CAD models. Bahubalendruni and Biswal attempted to find economic feasible assembly direction using bounding box coordinates and collision detection method [23], [24]. Although most of the researchers considered basic assembly predicates, stability and mechanical feasibility predicate were ignored due to the complex procedure of extraction [25]-[28]. The influence of ignoring these assembly predicates greatly impacts search space, computational time and quality of solution [29]. In the current research article, methods to extract all the assembly attributes required to perform optimal assembly sequences from CAD database are clearly illustrated.

II. Assembly Contact Information Extraction

An appending part must exhibit at least one contact with respect to any other part in the existent assembly subset. Graphical representation of assembly contact relations proposed by Bourjault [1] can be effectively used for this purpose. Dini further simplified the representation of assembly contact information in matrix mode for ease of accessibility. General method to extract and store liaison matrix in a two-dimensional array through assembly contact analysis from an assembled CAD product is represented in Fig. 1.

The presented method is implemented on assembly shown in Fig. 2, and the resulted liaison matrix is represented below.

B. B. Biswal is with the National Institute of Technology, Rourkela, Odisha -769008, India (corresponding author, phone: +91 9437158476; e-mail: bbbiswal@nitrkl.ac.in).

M. V. A. Raju Bahubalendruni, is with GMR Institute of Technology, Rajam India. (e-mail: bahubalindruni@gmail.com).

B. B. V. L. Deepak is with the National Institute of Technology, Rourkela, Odisha -769008, India (e-mail: bbv@nitrkl.ac.in).

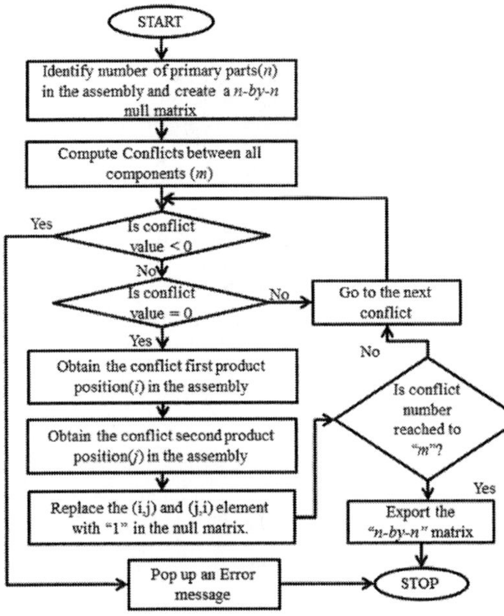

Fig. 1 Mechanism to extract liaison matrix from 3D CAD environment

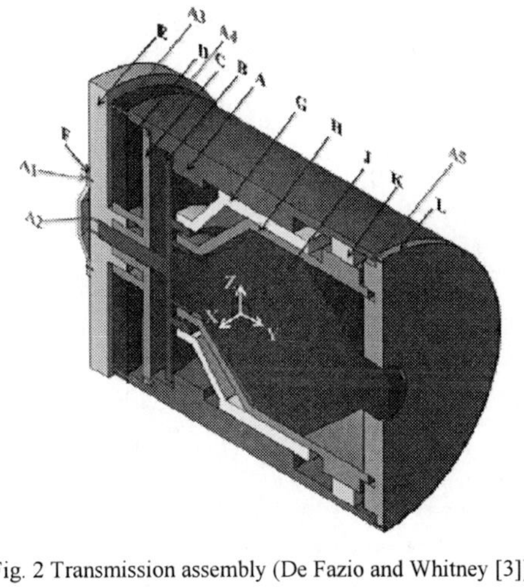

Fig. 2 Transmission assembly (De Fazio and Whitney [3])

	A	B	C	D	E	F	G	H	J	K	L
A	0	0	1	1	0	0	1	0	0	1	1
B	0	0	1	0	1	0	1	1	1	0	0
C	1	1	0	1	1	0	0	0	0	0	0
D	1	0	1	0	0	0	0	0	0	0	0
E	0	1	1	0	0	1	0	0	0	0	0
F	0	0	0	0	1	0	0	0	0	0	0
G	1	1	0	0	0	0	0	1	0	0	0
H	0	1	0	0	0	0	1	0	1	1	1
J	0	1	0	0	0	0	0	1	0	0	1
K	1	0	0	0	0	0	0	1	0	0	0
L	1	0	0	0	0	0	0	1	1	0	0

III. Geometrical Feasibility Information Extraction

In order to retrieve the assembly feasibility information, collision detection methods should be used efficiently considering part geometrical boundaries. Each part axis-aligned bounding box coordinates are considered to represent the geometric boundaries for the collision free part trajectories. Fig. 3 represents lower corner and upper corner of bounding box coordinates of a primary part (Part-C).

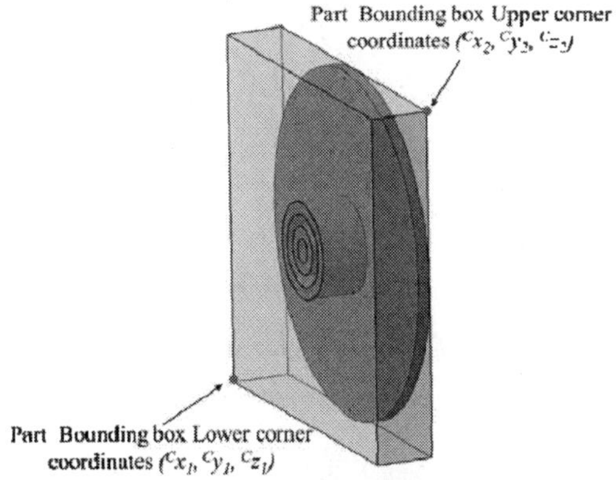

Fig. 3 Part-C Bounding box representation

There exist three different instances for any two parts (part-i and part-j). Instance-1: If the lower limit of part-i is more than the upper limit of part-j, and hence part-j does not interfere, while part-i is disassembled along the specified positive direction. These instances must be ignored from the feasibility testing by reading bounding box coordinates of the parts and unit value has be assigned to the directed interference matrix element.

Instance-2: The lower limit of part-i is less than the upper limit of part-j, and hence, part-j may interfere, while part-i is disassembling along the specified positive direction. Difference between the upper limit of part-j to the lower limit of part-i will be considered as distance to be travelled by part-i without any collision. These instances must be considered for the feasibility testing. Instance-3: If the upper limit of part-i is less than the lower limit of part-j, then part-i will be moved towards part-j till both values are matched, and Instance-3 will be turned out to instance-2 category. All the three possible instances are shown in Fig. 4.

An algorithm to extract interference matrix along a positive direction (y+) is shown in Fig. 5. The method uses bounding box coordinates of part in combination with assembly contact analysis. Most of the mechanical codes have capabilities to translate the parts in 3D-space and perform assembly contact analysis. The proposed method can be similarly applied for all other directions (X-, X+, Y-, Z-, Z+).

The method is implemented on an assembly shown in Fig. 2, and the resulted interference matrix is given below. Element value "1" means that the part indicated in the column is feasible to dis-assemble along the specified direction in the presence of part indicated in row.

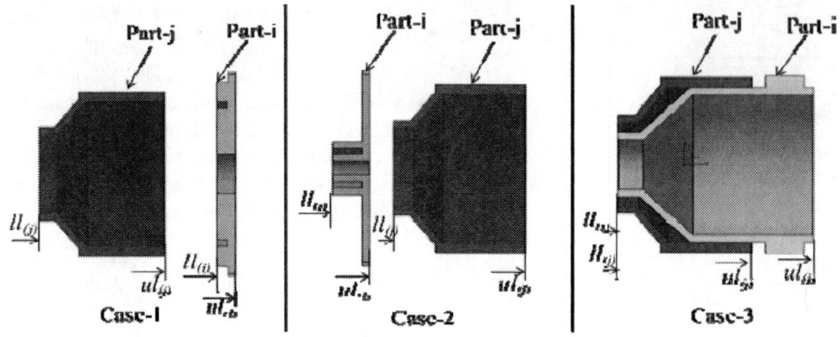

Fig. 4 Feasibility testing instances between pair of parts

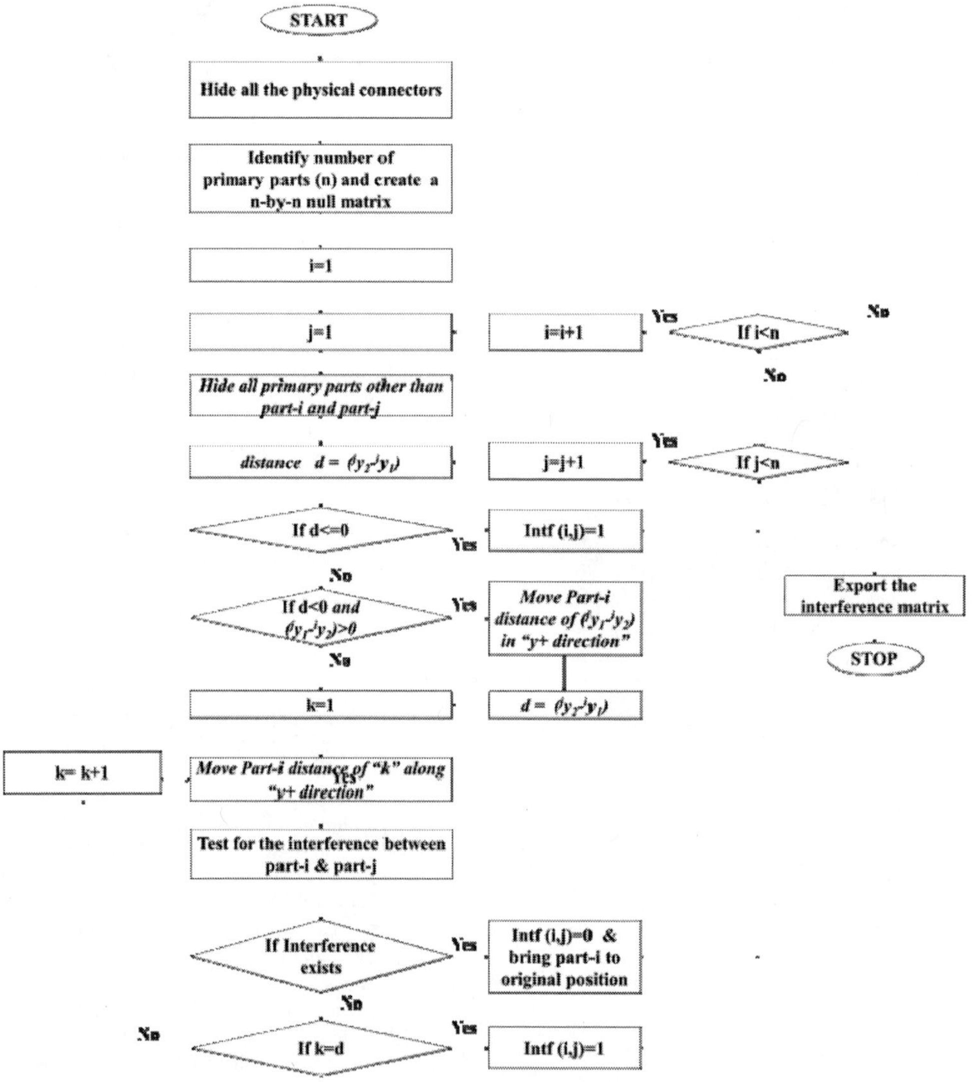

Fig. 5 Flow chart to extract Interference matrix along "y+" direction

(Y+)	A	B	C	D	E	F	G	H	J	K	L
A	0	1	1	1	1	1	0	0	1	0	0
B	0	0	1	1	1	1	0	0	0	0	0
C	0	0	0	1	1	1	0	0	0	0	0
D	0	0	0	0	1	1	0	0	0	0	0
E	0	0	0	0	0	1	0	0	0	0	0
F	1	0	0	0	0	0	0	0	0	1	0
G	1	1	1	1	1	1	0	0	0	1	0
H	1	1	1	1	1	1	1	0	0	1	0
J	1	1	1	1	1	1	1	1	0	1	0
K	1	1	1	1	1	1	1	1	1	0	0
L	1	1	1	1	1	1	1	1	1	1	0

IV. Stability Information Extraction

An assembly sequence is said to stable, when each of the part in the assembly maintain its position with respect to the other parts at all stages of assembly operations. Representation of stability using stable relations between each pair of parts is proposed by Smith et al. [30] and later modified by Bahubalendruni et al. [31]. Smith proposed representation of assembly connections in matrix format; the connections are categorized into two classes; hard and soft. When two parts are connected by physical connectors, the connection is considered as hard and if two components just maintain their position by surface contact without any physical connection is referred to be soft connection.

In the current research, stability is broadly classified as partial and permanent stability. A component is treated as partially stable when it does not lose its contact with all mating parts due to application of natural gravitational force. However, the assembly is oriented, and the parts may lose its contacts. Partial stability of part is more considered in sequential assembly planning process. Further classification of permanent stability is made by the usage of external attachments, and surface features. A component is treated as permanently stable, when it is connected through its surface features or by external connectors in order to maintain all its contacts with mating parts irrespective of the orientation. Permanent stability is an essential criterion for sub-assembly detection.

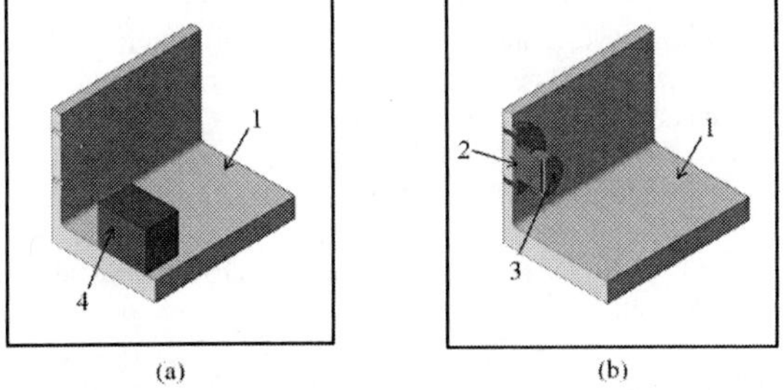

Fig. 6 Assembly sub-sets to demonstrate types of stability

Assembly subset (part 1-4) shown in Fig. 6 (a) is partially stable in which part-4 do not maintain its contact with respect to other mating part-1 when the assembly sub-set is rotated. Fig. 6 (b) represents a permanently stable assembly sub-set (1-2-3) which can be treated as a sub-assembly for the further level of assembly possess. The connection data for an assembled product can be represented by a "nxn" matrix for a product with "n" number of primary parts. Element [i][j] of the connection matrix represents how part-i is connected with part-j. Element values 0, 1, 2, and 3 successively represent no stability, partial stability, permanent stability due to part features and permanent stability by external physical connectors.

A. Partial Stability Matrix Extraction

Laws of equilibrium and stability of physical objects state: If a part cannot be rotated about an axis passing through its center of gravity parallel to the ground (XY- Plane). Fig. 7 briefs the possible instance of stability due to the resulted interferences while rotating the part in both clockwise ad counter clockwise orientations.

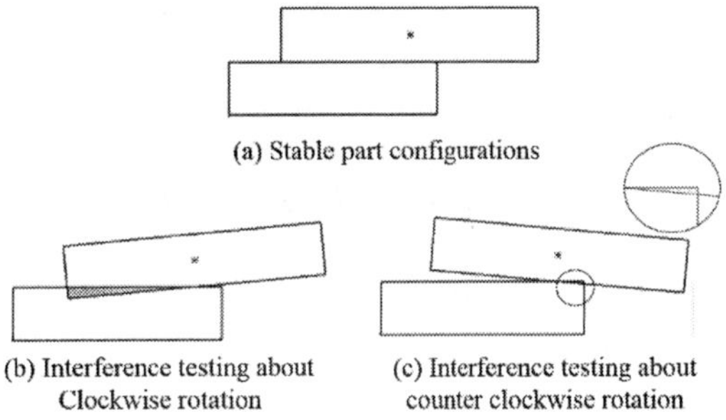

Fig. 7 Partially stable assembly sets

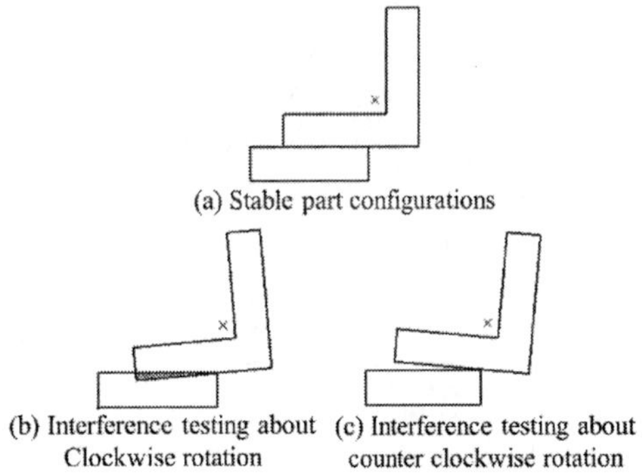

Fig. 8 Partially un-stable assembly sets

The possibility of instability due to clearance between the parts while rotated in counter clockwise orientation illustrated in Fig. 8. The contribution of liaison matrix and interference matrix about "z-" orientation will reduce the further efforts of partial stability matrix extraction.

Partial stability has to be checked for the pair of parts for which liaison matrix element and the interference matrix element values must be "1" and "0", respectively i.e indicating that the pair of parts are in contact and part-i (represented in row) cannot be disassembled along "z-" direction in the presence of part-j (represented in column). An algorithm to extract partial stability matrix is presented in the flow chart shown in Fig. 9.

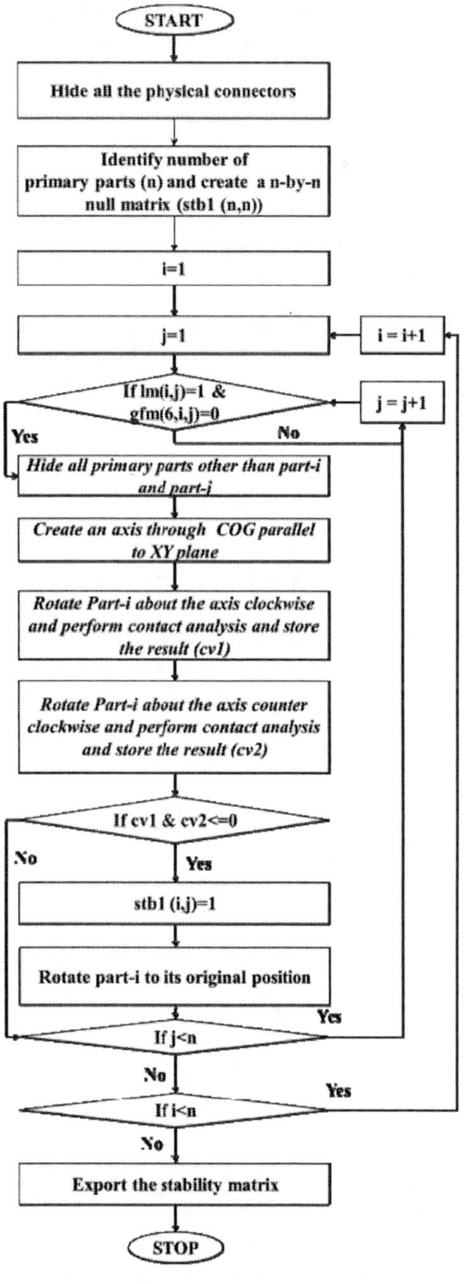

Fig. 9 Algorithm to extract partial stability matrix

The method is implemented on an assembly shown in Fig. 2, and the resulted partial stability matrix is given below.

	A	B	C	D	E	F	G	H	J	K	L
A	0	0	1	0	0	0	1	0	0	1	1
B	0	0	1	0	1	0	1	1	1	0	0
C	1	1	0	1	1	0	0	0	0	0	0
D	0	0	1	0	0	0	0	0	0	0	0
E	0	1	1	0	0	0	0	0	0	0	0
F	0	0	0	0	0	0	0	0	0	0	0
G	1	1	0	0	0	0	0	1	0	0	0
H	0	1	0	0	0	0	1	0	1	1	1
J	0	1	0	0	0	0	0	1	0	0	1
K	1	0	0	0	0	0	0	1	0	0	0
L	1	0	0	0	0	0	0	1	1	0	0

B. Permanent Stability Matrix Extraction (Due to Mating Features)

The part surface feature recognition is mainly dependent on the modelling methodology and the software interface compatibility to retrieve the information. Most of the advanced CAD codes (CATIA V5, SolidWorks, Pro E, Unigraphics, etc.,) offer flexibility to users in feature modelling and data extraction. Parts connected by their surface features such as threading, generally possess only one degree of freedom for assembly or disassembly operation. Hence, geometric feasibility matrices and liaison matrix data further minimize the complexity and computation time in retrieving the permanent stability information due to mating features. The method involves in extracting the part feature information (for example internal thread with a defined diameter and pitch on a lateral face) and tests for counter data on its mating part (External threading with same diameter and pitch) at the mating surface. A mechanism to extract permanent stability matrix is presented through a flowchart in Fig. 10.

Macro is written in VB script to get the permanent stable pair of parts, and permanent stability is indicated by "2" in the stability matrix. An assembly subset with four parts and four attachments cutaway shown in Fig. 11 is considered to demonstrate the permanent stability due to the part features.

For the assembly shown in Fig. 11, part-2 and part-3 are connected by means of surface threading and exhibits permanent stability. The resulted assembly stability matrix is given below.

C. Permanent Stability Matrix Extraction (Due to Physical Connectors)

Extraction of permanent stability matrix involved in identifying the connectors in the assembly model by their nomenclature and obtaining the pair of primary parts joined by the connector. Assembly contact analysis for a connector against all primary parts results pair of part those were connected by it, iterating the process for each connector resulting in permanent stability matrix. A computer aided method to extract permanent stability matrix is presented in Fig. 12 through flow chart.

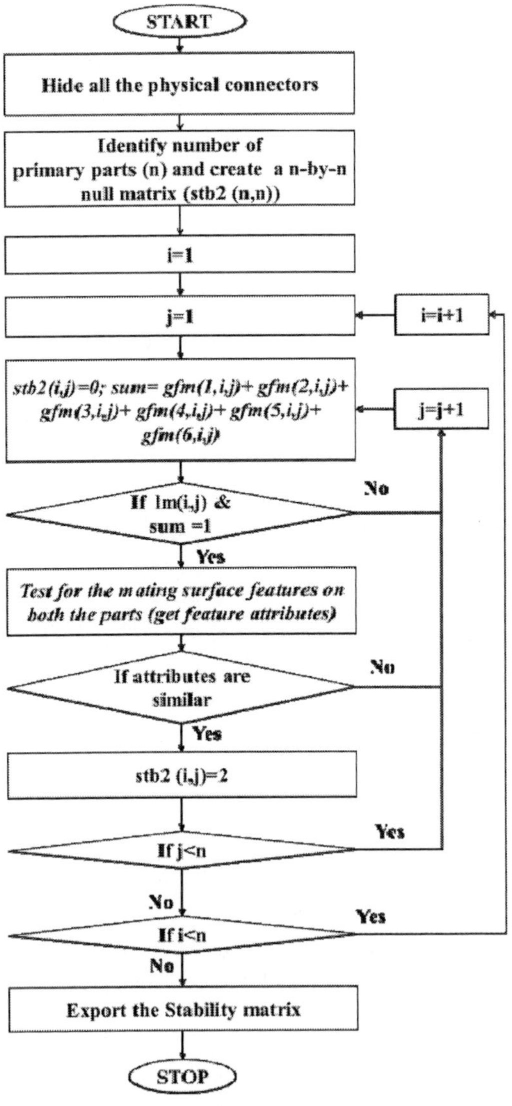

Fig. 10 Permanent stability matrix extraction method

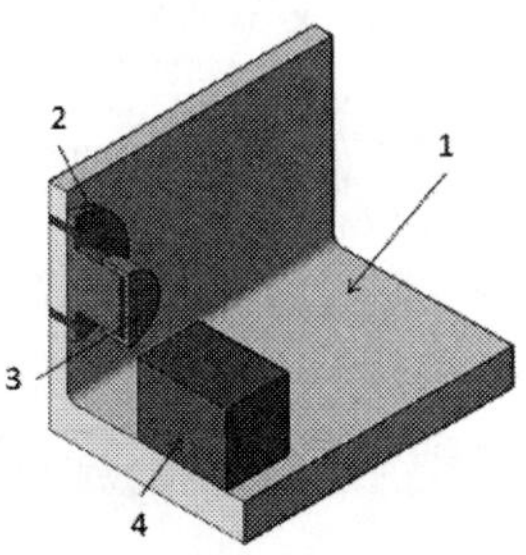

Fig. 11 Cutaway of 4-Part Assembly subset with 4 attachments

	1	2	3	4
1	0	0	0	0
2	0	0	2	0
3	0	2	0	0
4	0	0	0	0

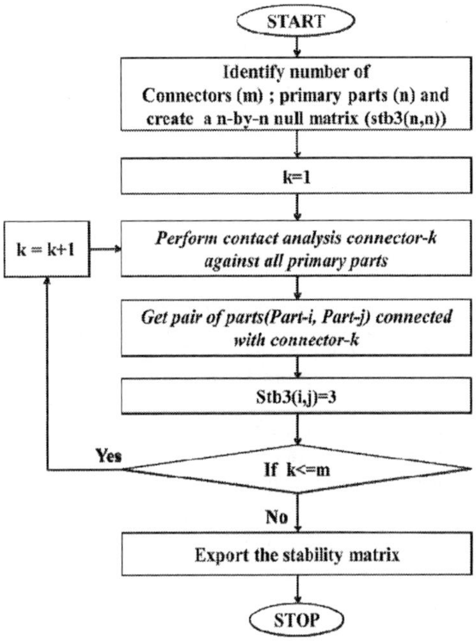

Fig. 12 Cutaway of four-part assembly subset with four attachments

Permanent stability matrix due to external connectors is extracted through a program written in VB script to interface with CATIA v5 environment. The resulted matrix is given below for the assembled product which is shown in Fig. 2. Permanent Stability matrix for transmission assembly due to external connectors is presented as:

	A	B	C	D	E	F	G	H	J	K	L
A	0	0	3	3	0	0	0	0	0	0	3
B	0	0	0	0	3	0	0	0	0	0	0
C	3	0	0	0	0	0	0	0	0	0	0
D	3	0	0	0	0	0	0	0	0	0	0
E	0	3	0	0	0	3	0	0	0	0	0
F	0	0	0	0	3	0	0	0	0	0	0
G	0	0	0	0	0	0	0	0	0	0	0
H	0	0	0	0	0	0	0	0	0	0	0
J	0	0	0	0	0	0	0	0	0	0	0
K	0	0	0	0	0	0	0	0	0	0	0
L	3	0	0	0	0	0	0	0	0	0	0

The combined stability matrix can be obtained by choosing the highest element value from partial and permanent stability matrices due to the possibility of conversion from partial stability condition to permanent stability condition due to external connectors. Combined stability matrix for transmission assembly shown in Fig. 2 is given as:

$$
\begin{array}{c} \\ A \\ B \\ C \\ D \\ E \\ F \\ G \\ H \\ J \\ K \\ L \end{array}
\begin{array}{c}
\begin{array}{ccccccccccc} A & B & C & D & E & F & G & H & J & K & L \end{array} \\
\begin{bmatrix}
0 & 0 & 3 & 3 & 0 & 0 & 1 & 0 & 0 & 1 & 3 \\
0 & 0 & 1 & 0 & 3 & 0 & 1 & 1 & 1 & 0 & 0 \\
3 & 1 & 0 & 1 & 1 & 0 & 0 & 0 & 0 & 0 & 0 \\
3 & 0 & 1 & 0 & 0 & 0 & 0 & 0 & 0 & 0 & 0 \\
0 & 3 & 1 & 0 & 0 & 3 & 0 & 0 & 0 & 0 & 0 \\
0 & 0 & 0 & 0 & 3 & 0 & 0 & 0 & 0 & 0 & 0 \\
1 & 1 & 0 & 0 & 0 & 0 & 0 & 1 & 0 & 0 & 0 \\
0 & 1 & 0 & 0 & 0 & 0 & 1 & 0 & 1 & 1 & 1 \\
0 & 1 & 0 & 0 & 0 & 0 & 0 & 1 & 0 & 0 & 1 \\
1 & 0 & 0 & 0 & 0 & 0 & 0 & 1 & 0 & 0 & 0 \\
3 & 0 & 0 & 0 & 0 & 0 & 0 & 1 & 1 & 0 & 0
\end{bmatrix}
\end{array}
$$

V. Mechanical Feasibility Information Extraction

In order to test the mechanical feasibility, firstly number of connectors must be identified from the stability matrix. By using the stability matrix, pair of parts with element value "3" must be acknowledged and contact analysis should be done for both the parts against all the physical connectors [32]. The common connectors are grouped into one set. While placing set of connectors, any part in the existed assembly other than those pair of parts can offer collision. Each connector will be tested for the geometric feasibility in the presence of other primary parts by using their bound box coordinates. Fig. 13 presents the flowchart to retrieve mechanical feasibility matrix using stability matrix data.

Mechanical feasibility matrix extraction method is presented in flow chart shown in Fig. 13, and a program is written in VB script to extract the mechanical feasibility matrix for the transmission assembly which is shown in Fig. 14.

VI. Mechanical Feasibility Information Extraction

In order to extract the assembly attribute information from CAD software through the above stated mechanisms, the software must be equipped with the basic part design and assembly design/representation module with several capabilities. Table I lists the basic requirements and their purpose in assembly attribute extraction. Table II lists the most used mechanical design software and their compatibility to extract the assembly attribute information.

VII. Conclusions

Detailed methods and computer aid to extract assembly attributes such as liaison matrix, geometrical feasibility matrices, mechanical feasibility matrices, and partial and permanent stability matrices are clearly illustrated with the example products. The proposed automated methods considering the laws of physical equilibrium along with the basic capabilities of CAD codes to retrieve assembly attributes are discussed. These methods are proven in extracting assembly attribute information without any skilled user-intervention. These methods greatly reduce the human errors and ease the process of generation assembly sequence plans. The captured assembly information is more precise and accurate; hence, the data can be directly used for the automated assembly sequence generation problems without any efforts.

TABLE I
BASIC REQUIREMENT OF CAD CODES FOR ASSEMBLY ATTRIBUTE EXTRACTION

S. No.	Assembly Attributes	Basic requirements of CAD software	Purpose
1	Liaison Matrix	Visualization filters	To hide/unhide connectors
		Assembly Clash check	To detect part contacts
2	Bounding Box Coordinates	Stock material measurement (or). STL conversion capability	To detect distances between parts along all principal axes (For interference checking)
3	Interference matrices	Part transformations	Part trajectories
		Assembly Clash check	Collision detection
		Visualization filters	To hide/unhide connectors and primary parts
4	Partial Stability matrix	Inertia Properties	COG detection
		3D-Rotation	Stability check
		Assembly Clash check	To detect part contacts
5	Permanent Stability matrix (Surface features)	Assembly feature's properties recognition	To detect similar features at contact faces.
6	Permanent Stability matrix (External connectors)	Part detection by nomenclature	Connectors identification
		Assembly Clash check	To detect part contacts
7	Mechanical feasibility Matrix	Part transformations	Part trajectories
		Assembly Clash check	Collision detection
		Visualization filters	To hide/unhide connectors and primary parts

TABLE II
MECHANICAL CAD CODES COMPATIBILITY FOR ASSEMBLY ATTRIBUTE EXTRACTION

S. No.	Software	Publisher	Direct Compatibility	API Compatibility
1.	Autodesk Inventor	AUTODESK INC.	3D solid Modelling and Assembly modelling.	C#
2.	Creo parametric (Pro Engineer)	Parametric Technology Corporation (PTC)	3D solid Modelling and Assembly modelling.	VB API/C++
3.	CATIA	Dassault Systems	3D solid Modelling and Assembly modelling.	CATScript /VBScript
4.	NX (UG or Unigraphics)	Siemens	3D solid Modelling and Assembly modelling.	C/C++
5.	Solidworks	Dassault Systems	3D solid Modelling and Assembly modelling.	Visual Basic

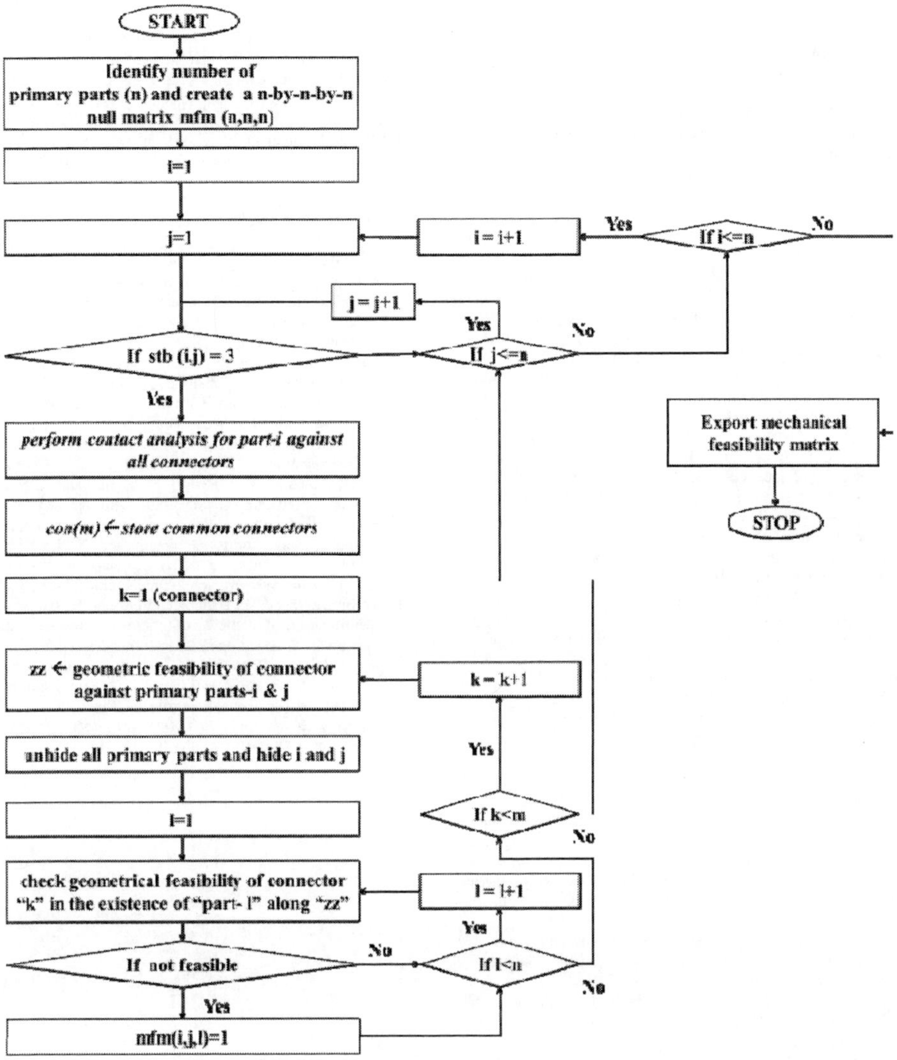

Fig. 13 Method to extract mechanical feasibility matrix

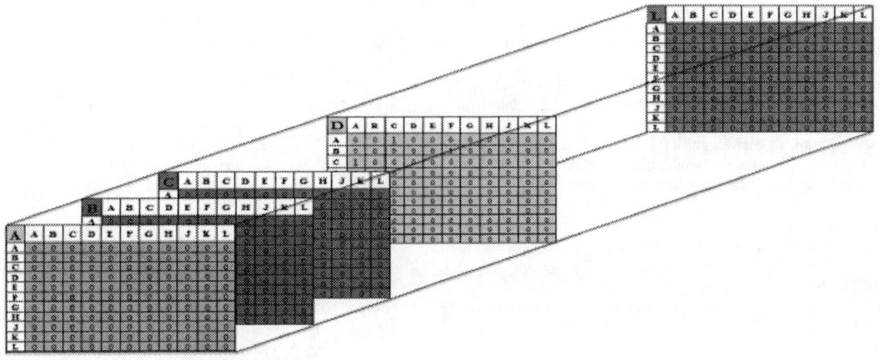

Fig. 14 Mechanical feasibility matrix for transmission assembly

REFERENCES

[1] A. Bourjault., 'Contribution a une approache methodologique de l'assemblage automatise: elaboration automatique des sequence s operatories', Ph.D. Thesis, L'Universite de Franche-Comte, 1984.

[2] L. Homem de Mello and A. Sanderson, "A correct and complete algorithm for the generation of mechanical assembly sequences", *IEEE Trans. Robot. Automat.*, vol. 7, no. 2, pp. 228-240, 1991.

[3] T. De Fazio and D. Whitney, "Simplified generation of all mechanical assembly sequences", IEEE J. Robot. Automat., vol. 3, no. 6, pp. 640-658, 1987.

[4] M. Bahubalendruni and B. Biswal, "A review on assembly sequence generation and its automation", *Proceedings of the Institution of Mechanical Engineers, Part C: Journal of Mechanical Engineering Science*, vol. 230, no. 5, pp. 824-838, 2015.

[5] M. R. Bahubalendruni, B. B. Biswal, and V. Upadhyaya, Assembly Sequence Generation and Automation. *In Proceeding of International Symposium on Engineering and Technology*, pp185-192, 2014.

[6] M. R. Bahubalendruni, B. Biswal, and G. R. Khanolkar, Computer aid to obtain assembly cut.-sets from 3D CAD product. *In International Conference on Design, Manufacturing and Mechatronics*, pp. 161-167, 2014.

[7] G. Dini, F. Failli, B. Lazzerini and F. Marcelloni, "Generation of Optimized Assembly Sequences Using Genetic Algorithms", *CIRP Annals - Manufacturing Technology*, vol. 48, no. 1, pp. 17-20, 1999.

[8] R. Nayak, M. Bahubalendruni, B. Biswal and M. Kumar, "Comparison of liaison concatenation method with simulated annealing for assembly sequence generation problems", *2015 1st International Conference on Next Generation Computing Technologies (NGCT)*, 2015.

[9] M. Bahubalendruni, B. Biswal and B. Deepak, "Optimal Robotic Assembly Sequence Generation Using Particle Swarm Optimization", *Journal of Automation and Control Engineering*, vol. 4, no. 2, pp. 89-95, 2016.

[10] M. Bahubalendruni, B. Deepak and B. Biswal, "An advanced immune based strategy to obtain an optimal feasible assembly sequence", *Assembly Automation*, vol. 36, no. 2, pp. 127-137, 2016.

[11] H. Shan, S. Zhou and Z. Sun, "Research on assembly sequence planning based on genetic simulated annealing algorithm and ant colony optimization algorithm", *Assembly Automation*, vol. 29, no. 3, pp. 249-256, 2009.

[12] C. Pan, S. Smith and G. Smith, "Automatic assembly sequence planning from STEP CAD files", *International Journal of Computer Integrated Manufacturing*, vol. 19, no. 8, pp. 775-783, 2006.

[13] A. Mathew and C. Rao, "A CAD system for extraction of mating features in an assembly", *Assembly Automation*, vol. 30, no. 2, pp. 142-146, 2010.

[14] R. Viganò and G. Osorio Gómez, "Assembly planning with automated retrieval of assembly sequences from CAD model information", *Assembly Automation*, vol. 32, no. 4, pp. 347-360, 2012.

[15] L. Ou and X. Xu, "Relationship matrix based automatic assembly sequence generation from a CAD model", *Computer-Aided Design*, vol. 45, no. 7, pp. 1053-1067, 2013.

[16] M. Bahubalendruni and B. Biswal, "Computer aid for automatic liaisons extraction from cad based robotic assembly", *2014 IEEE 8th International Conference on Intelligent Systems and Control (ISCO)*, 2014.

[17] Z. Ling. and S. Narayan, "Heuristic feature recognition from a CAD model". *Integrated Computer-Aided Engineering,* vol. 3, no. 3, pp. 212-224, 1996.

[18] T. Eng, Z. Ling, W. Olson and C. McLean, "Feature-based assembly modeling and sequence generation", *Computers & Industrial Engineering*, vol. 36, no. 1, pp. 17-33, 1999.

[19] M. R. Bahubalendruni, B. Biswal, and G. R. Khanolkar, "A Review on Graphical Assembly Sequence Representation Methods and Their Advancements ", *Journal of Mechatronics and Automation*, vol. 1, no.2, pp. 16-26, 2015.

[20] Alfadhlani, T. Samadhi, A. Ma'ruf and I. Toha, "Automatic Collision Detection for Assembly Sequence Planning Using a Three-Dimensional Solid Model", *Journal of Advanced Manufacturing Systems*, vol. 10, no. 02, pp. 277-291, 2011.

[21] P. GU and X. YAN, "CAD-directed automatic assembly sequence planning", *International Journal of Production Research*, vol. 33, no. 11, pp. 3069-3100, 1995.

[22] S. Mok, K. Ong and C. Wu, "Automatic generation of assembly instructions using STEP", *Proceedings 2001 ICRA. IEEE International Conference on Robotics and Automation (Cat. No.01CH37164).*

[23] M. Raju Bahubalendruni and B. Biswal, "An Intelligent Method to Test Feasibility Predicate for Robotic Assembly Sequence Generation", *Advances in Intelligent Systems and Computing*, pp. 277-283, 2014.

[24] M. Raju Bahubalendruni and B. Biswal, "An algorithm to test feasibility predicate for robotic assemblies." *Trends in Mechanical Engineering & Technology*. Vol.4, no.2 pp.11-16, 2014.

[25] W. Chen, Y. Hsu, L. Hsieh and P. Tai, "A systematic optimization approach for assembly sequence planning using Taguchi method, DOE, and BPNN", *Expert Systems with Applications*, vol. 37, no. 1, pp. 716-726, 2010.

[26] M. Raju Bahubalendruni and B. BISWAL, "Liaison concatenation – A method to obtain feasible assembly sequences from 3D-CAD product", *Sadhana*, vol. 41, no. 1, pp. 67-74, 2016.

[27] Q. Guan, J. Liu and Y. Zhong, "A concurrent hierarchical evolution approach to assembly process planning", *International Journal of Production Research*, vol. 40, no. 14, pp. 3357-3374, 2002.

[28] M. Bahubalendruni and B. Biswal, "A novel concatenation method for generating optimal robotic assembly sequences", *Proceedings of the Institution of Mechanical Engineers, Part C: Journal of Mechanical Engineering Science*, 2015. doi: 10.1177/0954406215623813.

[29] M. Bahubalendruni, B. Biswal, M. Kumar and R. Nayak, "Influence of assembly predicate consideration on optimal assembly sequence generation", *Assembly Automation*, vol. 35, no. 4, pp. 309-316, 2015.

[30] S. Shiang-Fong Smith, G. Smith and X. Liao, "Automatic stable assembly sequence generation and evaluation", *Journal of Manufacturing Systems*, vol. 20, no. 4, pp. 225-235, 2001.

[31] M. Kumar, M. V. A. Bahubalendruni and B. B. Biswal and R. Nayak, "Identification of stable configurations between constituent parts of an assembly", *International Conference on Mechanical Engineering Design (ICMED-2016), 25-26 April 2016*, Chennai, India, 2016.

[32] M. R. Bahubalendruni, B. B. Biswal, M. Kumar and B. B. V. L Deepak, "A Note on Mechanical Feasibility Predicate for Robotic Assembly Sequence Generation". *In CAD/CAM, Robotics and Factories of the Future* (pp. 397-404). Springer India, 2016.

RoboWeedSupport-Semi-Automated Unmanned Aerial System for Cost Efficient High Resolution in Sub-Millimeter Scale Acquisition of Weed Images

Simon L. Madsen, Mads Dyrmann, Morten S. Laursen, Rasmus N. Jørgensen

***Abstract*—**Recent advances in the Unmanned Aerial System (UAS) safety and perception systems enable safe low altitude autonomous terrain following flights recently demonstrated by the consumer DJI Mavic PRO and Phamtom 4 Pro drones. This paper presents the first prototype system utilizing this functionality in form of semi-automated UAS based collection of crop/weed images where the embedded perception system ensures a significantly safer and faster gathering of weed images with sub-millimeter resolution. The system is to be used when the weeds are at cotyledon stage and prior to the harvest recognizing the grass weed species, which cannot be discriminated at the cotyledon stage.

***Keywords*—**Weed mapping, integrated weed management, DJI SDK, automation, cotyledon plants.

I. Introduction

TODAY UAS (addressed as drone in the following) are being used worldwide for weed mapping. Patches of weed are easy to identify using centimeter pixel resolution and relatively simple algorithms [1]-[3]. However, to identify and classify weeds at the cotyledon stage using shape features [4]-[6], millimeter and sub-millimeter pixel resolution is needed [7], [8].

Jørgensen et al. [9] demonstrated off-the-shelf drone as a user-friendly platform for obtaining high-resolution (sub-millimeter ground sampling distance) images as input to automated weed recognition data processing. Weed mapping is conducted as low altitude (sub-meter) point sampling in field hot spots, such as headlands, depressions in the terrain, etc. However, Jørgensen et al. [9] point out that collecting images from sub-meter altitudes is not cost efficient, when using manual control of the drone, especially in uneven and undulated fields.

In contrast to classic drone flight planning with the purpose of making orthographic maps from >20 m altitudes, semi-automated weed mapping requires a special flight planning tool for controlling the drone. This tool should include an interface in which the user easily can point and create waypoints where the drone should collect images. Drone behavior, such as flight speed, image capture altitude, camera settings, etc. should be automated.

In this work, flight altitude is controlled by an active height sensor embedded on a consumer drone with no retrofits. The only required input from the pilot is to decide numbers and locations of waypoints, and to avoid obstacles in the flight plan due to the low altitude flying. The remaining flight control and image capturing are fully automated.

The system is expected to be used as a data collection platform in weed mapping systems, while the weeds are at the cotyledon stage [10]. Weed recognition and classification in the collected images will be handled by a separate software.

S. L. Madsen is with the Department of Engineering at Aarhus University, 8000 Aarhus C, Denmark (phone: +45 50409949; e-mail: slm@eng.au.dk).
M. Dyrmann is with the Maersk Mc-Kinney Moller Institut at University of Southern Denmark, 5230 Odense M, Denmark (e-mail: mady@mmmi.sdu.dk)
M. S. Laursen was with the Department of Engineering at Aarhus University, 8000 Aarhus C, Denmark (e-mail: msl@eng.au.dk).
R. N. Jørgensen is with the Department of Engineering at Aarhus University, 8000 Aarhus C, Denmark (e-mail: rnj@eng.au.dk).

II. Materials and Methods

The aim is to use a consumer drone with an active height sensor and Software Development Kit (SDK) availability. Experience from Madsen et al. [11] showed that the ultrasonic height sensor of the Phantom 4 (DJI, Shenzhen, China) drone was proven to be too unreliable to be used for actively adjusting the flight height. Instead, a DJI Matrice 100 (DJI, Shenzhen, China) retrofitted with a Velodyne VLP-6 LiDAR (Velodyne LiDAR, Morgan Hill, USA) had to be used. However, this solution is expensive and not suited for consumers. Hence, a DJI Phantom 4 Pro (DJI, Shenzhen, China) drone is chosen as case demonstrator in this study.

A. Mechanical Setup

The entire setup consists of an iPad Air (Apple Inc., California, USA) that communicates with a DJI Phantom 4 Pro drone trough the corresponding remote controller, see Fig. 1.

Fig. 1 Illustration of the iPad with running application and the DJI Phantom 4 Pro Drone after successful mission completion. Picture taken during test of the system on February 6, 2017

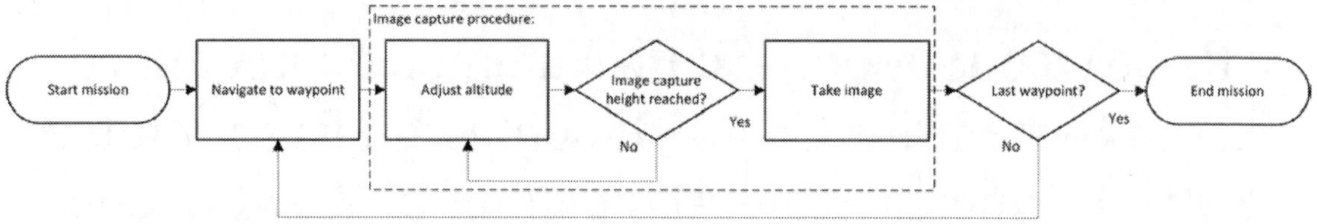

Fig. 2 State chart of mobile application during mission execution. The chart show how the application toggles between navigating to waypoint and the 'Image capture procedure'

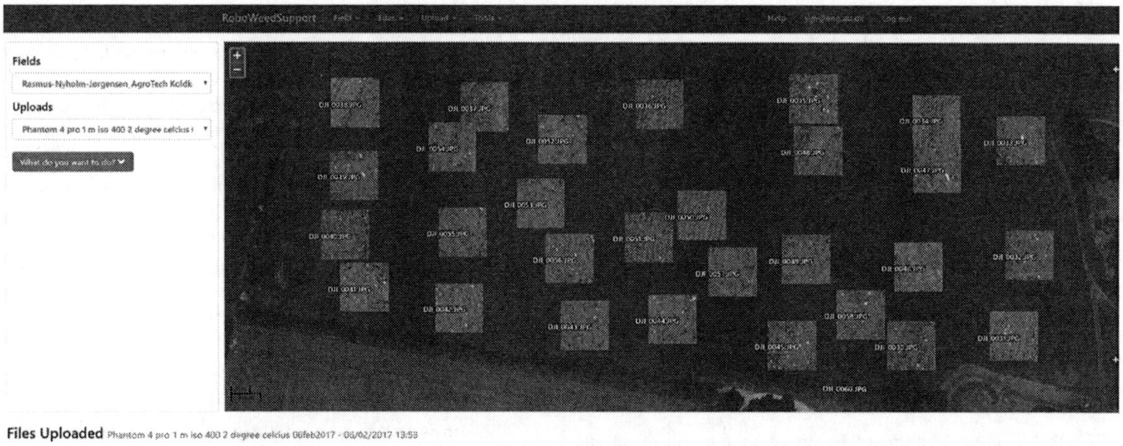

Fig. 3 Images acquired during test of the system performed February 6, 2017 and uploaded to the RoboWeedSupport cloud as described in details by Rydahl et al. [16] The images are plotted corresponding to the location they were taken

The Phantom 4 Pro drone is configured with the default accessory package, with exception of the battery that has been replaced by the Phantom 4 Series – Intelligent Flight Battery (5870 mAH, High Capacity) (DJI, Shenzhen, China), to ensure a longer flight duration. The Phantom 4 Pro drone is chosen as the platform, since it includes a better safety system and ultrasonic height sensor, compared to the older versions in the Phantom series. The improved ultrasonic height sensor provides a reliable height measure that is used throughout mission execution to make altitude adjustments.

In order to ensure sufficient resolution, the image acquisition height was set to 1 m. This results in an imaging area of ~1.12 m^2 and ~4 pixels mm^{-1}.

B. Mobile and Tablet Application

An iPad with iOS operating system is used as the application host. The iOS application handles route planning and is responsible for the image capture procedure. The iOS application is developed in Objective-C using Xcode (Apple Inc. California, USA) and the DJI mobile SDK (DJI, Shenzhen, China). The application is based on the DJI tutorial, 'GSDemo' [12] and is modified, so each time a waypoint is reached, the application starts a custom implemented image capture procedure. This procedure consists of two processes: 'Adjust altitude' and 'Take image. The flow of the waypoint mission is shown in Fig. 2.

The waypoint mission component does not incorporate the height measures from the active height sensor on the Phantom 4 Pro and is therefore not suited nor able to control the drone close to the ground (safety system interferes at 2 m above ground and below). Instead, the 'Adjust altitude' process stops the waypoint mission temporarily, and the iPad application takes direct control of the drone through the virtual joystick component. The process is implemented as a control loop that uses the ultrasonic sensor of the Phantom 4 Pro to adjust the vertical velocity of the drone with an update frequency of 10 Hz. The adjustment is implemented as a linear function of the difference in current and desired height, so the drone deaccelerates as it comes closer to the image capture height.

When the drone is within ±0.1 m of the desired image capture height, the 'Take image' process is executed. The process will take an image of the ground and will resume the waypoint mission. When resuming the mission, the flight and image capture altitude are adjusted to compensate for height variations in the field.

III. Results

Fig. 3 shows the images acquired during a test of the system performed February 6, 2017 at 1:50 pm. under cloudy and relatively dull conditions and fixed ISO (= 400) resulting in exposure times of 1/250 to 1/120 second. The test flight was performed in a winter wheat field in eastern Jutland, Denmark

(56°12'11.8"N, 10°09'01.8"E). The images are plotted on a map to show the location in the field where they are captured. Because of the low altitude, the down force from the drone rotors created motion blur in the images, in particular for the winter wheat leave tips. This was not the case in most of the weed cases due to their close to ground posture.

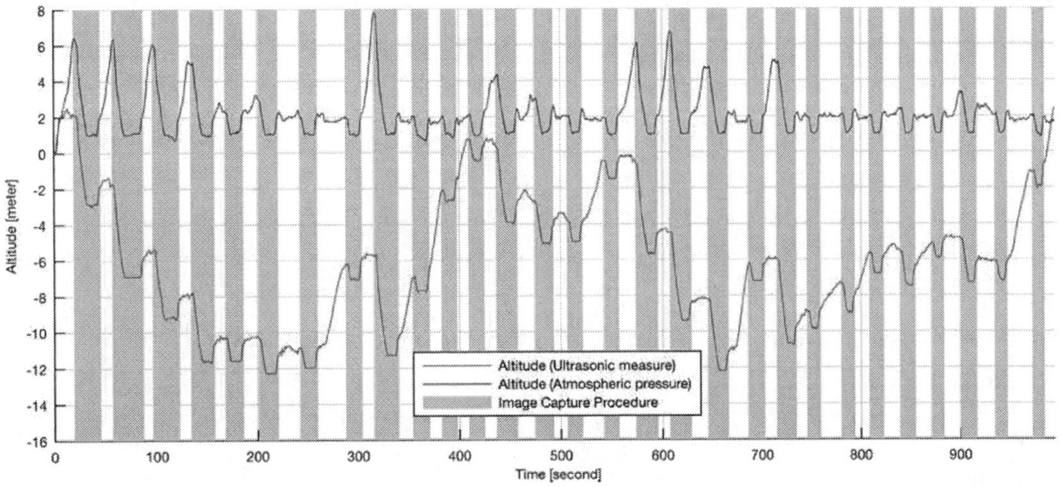

Fig. 4 Recorded altitude measures of the drone during test flight February 6, 2017. The ultrasonic altitude shows the operational flight height relative to the ground throughout mission execution. The atmospheric pressure altitude shows the flight height relative to the home position and gives an indication of the variations in the terrain. The areas highlighted in gray indicate the time used in the image capture procedure, whereas the remainder is used on take-off, navigation between waypoints and end mission action. The plot shows how the software adapts the flight and capture altitude to height differences in the field

From pressing "start mission" within the app shown in Fig. 1 on the tablet until the Phantom 4 Pro landed again, the drone operator did not touch the remote control or intervened in any way until the drone landed again 17 minutes later. This was a huge relief compared to manually collecting sub-meter images with a DJI Phantom 4 as demonstrated by Jørgensen et al. [9].

Table I summarizes key numbers from the test flight on February 6, 2017. The numbers show that the system has a capacity of ~2 images min^{-1}, with an image density of ~5 image ha^{-1}. The table shows that the image capture procedure is highly time consuming, which is also apparent by plotting the recorded altitude of the drone throughout the test flight, see Fig. 4.

TABLE I
KEY NUMBERS FROM TEST OF THE SYSTEM PERFORMED ON THE FEBRUARY 6, 2017

Parameter	Value
Number of images	29
Set Flight altitude	2.2 m
Set Image capture altitude	1 m
Total flight time	1028.1 s
Area covered	~ 6 ha
Distance traveled	~ 1900 m
Mean time per image (navigation to waypoint + image capture procedure)	34.0 ± 4.9 s
Mean time image capture procedure	18.4 ± 4.5 s

The 29 images were analyzed through manually annotation and automated software for detection of weeds as presented by Dyrmann et al. [13]. The manual annotation was able to register 851 occurrences of weeds, whereas the automated software was able to register 329 occurrences from the images. An example of the results from the automated weed detection is shown in Fig. 5.

The detected weeds were also attempted classified using a modified version of the automated classifications software presented by Dyrmann et al. [5]. However, the classification did not provide any useable results, since the quality of images is too low, due to the poor weather conditions.

IV. Discussion

This study has demonstrated that consumer drones are capable of being used for semi-autonomous image acquisition with resolution in sub-millimeter scale. However, experience from the test of the system has shown two issues that will need to be handled in future versions of the system: The image quality, and the overall capacity of the system with respect to area covered versus time spent.

The quality of the images recorded during test of the system on February 6, 2017 is sufficient for detection of weeds, but is not good enough for classification of the weeds. The poor image quality is probably a result of the cloudy and relatively dull weather conditions. The experienced image blur could also be an effect of the relatively high ISO resulting in the camera grouping pixels together to capture more light. This issue should be relatively easy to fix, since it only requires better planning with regard to the weather and other camera settings on the drone.

Fig. 5 Results from using the automated weed detection software presented by Dyrmann [13]. The red squares show the locations where the weeds are detected. The figure also shows that the detection is not perfect, since it fails at detecting some weeds

It is not surprising that approximately half of the flight time is used on the image capture procedure as seen in Fig. 4 and Table I, since the procedure is not fully optimized. Pausing the waypoint mission, descending and ascending the drone and resuming the mission are all very time consuming, since it depends on off-drone remote control on the iOS-based iPad. If this procedure was embedded in the drone's flight controller as part of the DJI SDK, the time could probably be reduced. Still the capacity of the drone-based system in this work is rather low compared to the ATV based weed mapping solution described by Laursen et al. [14]. Assuming the drone-based system manages to decrease the time of the image capture procedure by a factor two and the flight time is increased to 30 minutes (maximum flight time listed by DJI [15]), then approximately 73 images of 1.12 m^2 equal to ~82 m^2 of the field can be imaged in one flight. The image area of the ATV system presented by Laursen et al. [14] is approximately quarter compared to solution presented in this work, but ~1400 images are collected in half an hour covering 14 hectares resulting in 350 m^2 of the field imaged. The ATV solution is estimated to be a 20,000 EUR investment, whereas the solution presented in this study is estimated to approximately 2,200 EUR. This leaves plenty of room for investing in several consumer drones and reaching the same or higher capacity than the ATV solution. In addition, the drone does not create tracks in the fields. Assuming that future low cost consumer drones can fulfill the latter capacity improvements, it will be a valuable tool in agriculture if linked up with analytics systems, e.g. the RoboWeedSupport cloud system described by Rydahl et al. [16].

The results presented by Madsen et al. [11] show that it could be beneficial to use the latest Normalized Difference Vegetation Index (NDVI) map as a reference, when planning the spatial distribution of the image acquisition points prior to flight. This could be implemented by adding the NDVI as a layer on top of the map in the iOS application. Since the Sentinel-2 and Landsat satellite imagery is free and relatively easy to access, e.g. ESA Scientific Hub, Amazon AWS, Google Cloud Platform [17], [18], this feature will be implemented in future version of the software developed in this work.

V. Conclusions

In conclusion, consumer drones like the DJI Phantom 4 Pro can be used for autonomous image acquisition with sub-millimeter pixel resolution. However, there are rooms for improvements for the proposed solution, such as further increasing the quality of the acquired images and optimizing the flight control, in order to increase the capacity of the system. In order to become a truly valuable weed mapping tool, the system should ideally be linked with fully automated weed recognition software and integrated weed management decision support systems.

ACKNOWLEDGMENTS

This work was founded by 1) GUDP (Grønt Udviklings- og Demonstrations Program) under the Ministry of Environment and Food of Denmark via the RoboWeedSupport project; and 2) Innovation Fund Denmark via the RoboWeedMaPS project.

REFERENCES

[1] J. M. Peña, J. Torres-Sánchez, A. I. de Castro, M. Kelly, and F. López-Granados, "Weed mapping in early-season maize fields using object-based analysis of unmanned aerial vehicle (UAV) images," *PLoS One*, vol. 8, no. 10, p. e77151, 2013.

[2] D. Gómez-Candón, A. I. De Castro, and F. López-Granados, "Assessing the accuracy of mosaics from unmanned aerial vehicle (UAV) imagery for precision agriculture purposes in wheat," *Precis. Agric.*, vol. 15, no. 1, pp. 44–56, 2013.

[3] J. M. Peña, J. Torres-Sánchez, A. Serrano-Pérez, A. I. de Castro, and F. López-Granados, "Quantifying Efficacy and Limits of Unmanned Aerial Vehicle (UAV) Technology for Weed Seedling Detection as Affected by Sensor Resolution," *Sensors*, vol. 15, no. 3, pp. 5609–5626, 2015.

[4] M. Dyrmann and R. N. Jørgensen, "RoboWeedSupport: weed recognition for reduction of herbicide consumption," in *Precision agriculture '15*, J. V Stafford, Ed. Wageningen Academic Publishers Books, 2015, pp. 571–578.

[5] M. Dyrmann, H. Karstoft, and H. S. Midtiby, "Plant species classification using deep convolutional neural network," *Biosyst. Eng.*, vol. 151, pp. 72–80, Nov. 2016.

[6] M. Laursen, R. Jørgensen, H. Midtiby, K. Jensen, M. Christiansen, T. Giselsson, A. Mortensen, and P. Jensen, "Dicotyledon Weed Quantification Algorithm for Selective Herbicide Application in Maize Crops," *Sensors*, vol. 16, no. 11, p. 1848, 2016.

[7] R. B. Brown and S. D. Noble, "Site-specific weed management: sensing requirements--- what do we need to see?," *Weed Sci.*, vol. 53, no. 2, pp. 252–258, 2005.

[8] D. E. Guyer, G. E. Miles, M. M. Schreiber, O. R. Mitchell, and V. C. Vanderbilt, "Machine vision and image processing for plant identification," *Trans. ASAE*, vol. 29, no. 6, pp. 1500–1507, 1986.

[9] R. N. Jørgensen, M. S. Laursen, M. Dymann, and R. N. Poulsen, "RoboWeedSupport - Weed Mapping with drones using a DJI Phantom 4," 2016. (Online). Available: https://www.youtube.com/watch?v=yegTo2mw6GA. (Accessed: 07-Feb-2017).

[10] J. Rasmussen, J Nielsen, F Garcia-Ruiz, and J. C. Streibig, "Potential uses of small unmanned aircraft systems (UAS) in weed research," *Weed Res.*, vol. 53(4), no. 242–248, pp. 242–248, 2013.

[11] S. L. Madsen, M. S. Larsen, R. N. Poulsen, and R. N. Jørgensen, "RoboWeedSupport - Semi-automated UAS system for cost efficient high resolution in sub-millimeter scale acquisition of weed images," in *ECPA 2017 - 11th European Conference on Precision Agriculture*, 2017.

[12] L. DJI Technology CO., "Creating a MapView and Waypoint Application," *DJI Mobile SDK Documentation*, 2016. (Online). Available: https://developer.dji.com/mobile-sdk/documentation/ios-tutorials/GSDemo.html. (Accessed: 07-Feb-2017).

[13] M. Dyrmann, R. N. Jørgensen, and H. S. Midtiby, "RoboWeedSupport - Detection of Weed Locations in Leaf Occluded Cereal Crops using a Fully Convolutional Neural Network," in *ECPA 2017 - 11th European Conference on Precision Agriculture*, 2017.

[14] M. S. Laursen, R. N. Jørgensen, M. Dyrmann, and R. N. Poulsen, "RoboWeedSupport - Sub millimeter weed image acquisition in cereal crops with speeds up till 50 km/h," in *ICPA 2017 - 19th International Conference on Precision Agriculture*, 2017.

[15] L. DJI Technology CO., "Phantom 4 Pro Specs," 2017. (Online). Available: https://www.dji.com/phantom-4-pro/info. (Accessed: 07-Feb-2017).

[16] P. Rydahl, N.-P. Jensen, M. Dyrmann, P. H. Nielsen, and R. N. Jørgensen, "RoboWeedSupport - Presentation of a cloud based system bridging the gap between in-field weed inspections and decision support systems," in *ECPA 2017 - 11th European Conference on Precision Agriculture*, 2017.

[17] J. H. Jeppesen, R. H. Jacobsen, R. N. Jørgensen, R. Gislum, A. Halberg, and S. T. Toftegaard, "Improving Profitability in Precision Agriculture by Identification of High-Variation Fields based on Open Satellite Imagery," in *ECPA 2017 - 11th European Conference on Precision Agriculture*, 2017.

[18] J. H. Jeppesen, R. H. Jacobsen, R. Nyholm Jørgensen, and T. S. Toftegaard, "Towards Data-Driven Precision Agriculture using Open Data and Open Source Software," in *International Conference on Agricultural Engineering 2016*, 2016.

World Academy of Science, Engineering and Technology
International Journal of Chemical, Molecular, Nuclear, Materials and Metallurgical Engineering Vol:11, No:5, 2017

International Science Index, Materials and Metallurgical Engineering Vol:11, No:5, 2017 waset.org/Publication/10007167

A Review on Application of Phase Change Materials in Textiles Finishing

Mazyar Ahrari, Ramin Khajavi, Mehdi Kamali Dolatabadi, Tayebeh Toliyat, Abosaeed Rashidi

Abstract**—**Fabric as the first and most common layer that is in permanent contact with human skin is a very good interface to provide coverage, as well as heat and cold insulation. Phase change materials (PCMs) are organic and inorganic compounds which have the capability of absorbing and releasing noticeable amounts of latent heat during phase transitions between solid and liquid phases at a low temperature range. PCMs come across phase changes (liquid-solid and solid-liquid transitions) during absorbing and releasing thermal heat; so, in order to use them for a long time, they should have been encapsulated in polymeric shells, so-called microcapsules. Microencapsulation and nanoencapsulation methods have been developed in order to reduce the reactivity of a PCM with outside environment, promoting the ease of handling, decreasing the diffusion and evaporation rates. Methods of incorporation of PCMs in textiles such as electrospinning and determining thermal properties had been summarized. Paraffin waxes catch a lot of attention due to their high thermal storage density, repeatability of phase change, thermal stability, small volume change during phase transition, chemical stability, non-toxicity, non-flammability, non-corrosive and low cost and they seem to play a key role in confronting with climate change and global warming. In this article, we aimed to review the researches concentrating on the characteristics of PCMs and new materials and methods of microencapsulation.

Keywords**—**Thermoregulation, phase change materials, microencapsulation, thermal energy storage, nanoencapsulation.

I. Introduction

By changing the temperature of different parts of the body, as well as the environment temperature, textiles containing PCMs get affected and react proportional to the temperature of the ambient. Due to change of the physical state of PCMs (liquid-solid, solid-solid) in terms of different thermal conditions, application of the microencapsulation technique is necessary and requisite for preserve, protect and optimal use of PCM. The main role of the clothing is to protect the body from various thermal fluctuations and therefore use of microcapsules containing PCM (micro PCM) in manufacturing and finishing processes of textiles will significantly improve and enhance their thermoregulating property. Microencapsulated phase change materials (mPCMs) during absorbing and storing thermal energy and releasing of it come across solid-liquid and liquid-solid transitions, respectively. The mentioned process is called phase transition (phase change). PCMs store thermal energy during heating process and they release the previously stored thermal energy during cooling process and therefore they are widely being applied in manufacturing and finishing of various textiles. They can be directly applied within fibers (acrylic fiber) and foams (polyurethane foam) or as a lining in production of various fabrics. Naturally, the body has the ability of adapting itself with surrounding temperature changes in different climatic conditions. The majority of this adaptation is done by the skin, although abnormal thermal fluctuations can be fatal as actions such as shaking, horripilation, mouth and nose secretions, severe sweating and too much breathing are not able to cause automatic thermoregulation and performing intense physical exercise, application of proper clothing and artificial heating and cooling are inevitable for maintaining thermal equilibrium [1]. The technology of embedding PCMs inside fabric structure in order to enhance their thermal properties in the late 1980s had been taken under account via NASA's development program. At the beginning, the main goal of applying this technology was astronauts' space suits in order to provide higher protection against high thermal fluctuations in outer space [2]. PCMs were embedded into textile materials with the development of microencapsulation technologies in 1987. Currently, more than 500 PCMs are known which were embedded directly inside fibers either foams or applied topically. Among the various PCMs, paraffin waxes (n-paraffin) with a melting point between 18°C and 36°C can be mentioned (heptadecane, hexadecane, octadecane, nonadecane and eicosane). Paraffin waxes, according to the number of carbon atoms in their chemical structure, have various melting and crystallization temperatures. Paraffin waxes cannot be directly embedded into textiles as their melting point is low, so in order to embed them inside textiles, microencapsulation is inevitable. We can encapsulate PCMs inside fine polymeric flexible layers (microcapsules) via microencapsulation technique in which they can easily change from solid to liquid phase and vice versa [1].

M. Ahrari is with Department of Textiles Engineering, Sciences and Research Branch Islamic Azad University, Tehran, Iran (e-mail: ahrari.mazyar@gmail.com).
R. Khajavi is with Department of Textiles Engineering, South Tehran Branch, Islamic Azad University, Tehran, Iran.
M. Kamali Dolatabadi and A. Rashidi are with Department of Textiles Engineering, Sciences and Research Branch Islamic Azad University, Tehran, Iran.
T. Toliyat is with Department of Pharmaceutics, Tehran University of Medical Sciences, Tehran, Iran.

II. Researches and Studies in Relation to Application of PCMs in Textiles Finishing

Right now manufacturing heat storing and thermoregulating textiles and clothing include the following methods: electrospinning of fibers via various PCMs, addition of micro PCMs to fibers, textiles and foams and direct addition of PCMs to textiles, fibers and foams.

A. Electrospinning of Fibers via Various PCMs

A wide variety of thermoregulating fibers could not be manufactured via conventional spinning methods as the processability of many polymers and PCMs is difficult. The application of electrospinning is one of the proper and optimal methods of manufacturing thermoregulating fibers. In order to produce large quantities of ultra-fine fibers in micro and nanoscale containing a variety of polymers (polymeric mixtures and polymer impregnated with nanoparticles), electrospinning is a very simple, easy and flexible method. In electrospinning process (Fig. 1), mainly polymeric solutions are being applied. In order to manufacture nanofibers with a full range of composition, a variety of morphologies and properties, composites, surfactant based solutions and sol-gels are also used [3].

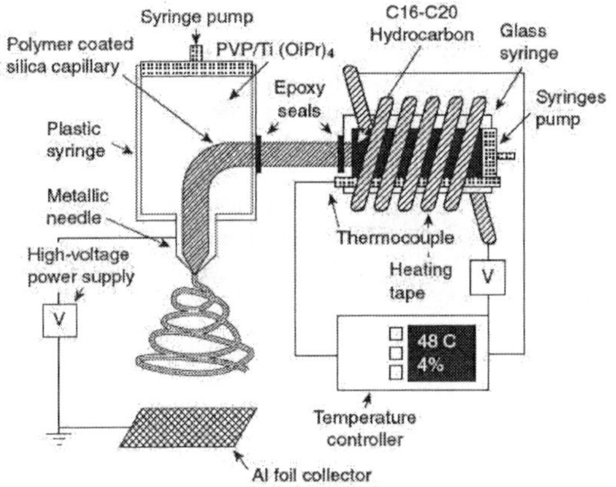

Fig. 1 Manufacturing of TiO_2-PVP nanofibers loaded with PCMs via melt coaxial electrospinning [4]

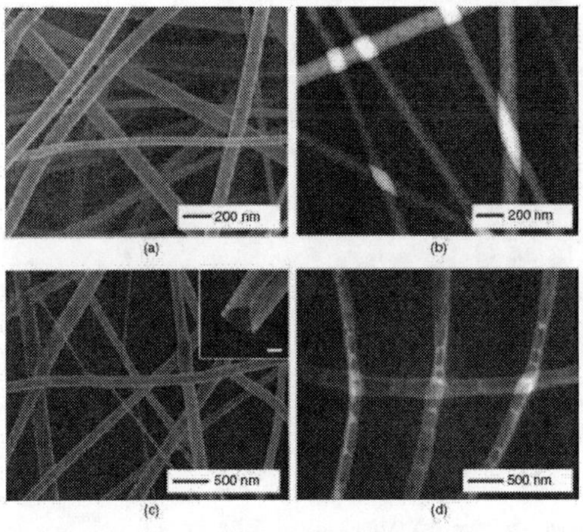

Fig. 2 Octadecane @ TiO_2-PVP nanofibers obtained via melt coaxial electrospinning with spinneret: (a) SEM micrograph of nanofibers with 7% octadecane, (b) TEM of nanofibers after removal of octadecane, (c) SEM micrograph of nanofibers with 45% octadecane and (d) TEM of nanofibers and octadecane removal [4]

Fig. 3 Electrospun fibers SEM micrographs: (a) CA fibers, (b) PEG/CA fibers, (c) washed PEG/CA fibers and (d) PEG/CA fibers after 100 heating and cooling cycles [5]

Melt coaxial electrospinning method for manufacturing phase change nanofibers consisted of n-octadecane and n-eicosane cores and composite polymeric shells had been developed in one step [4]. In the mentioned process, n-octadecane PCM was removed by impregnation in hexane for 24 hours. Obtained TiO_2-PVP (polyvinylpyrrolidone) nanoparticles due to the rapid freezing of hydrocarbons obtained from the cooling of the carrier solution's vapor, displayed novel segmentized morphologies. N-hexadecane-PVP, n-octadecane-PVP and n-eicosane-PVP nanofibers enthalpy changes at 16°C, 27°C and 37°C were 71 KJ/KG, 114 KJ/KG and 88 KJ/KG, respectively (Fig. 2).

Taking advantage of the electrospinning method, PEG 10000 and cellulose acetate (CA) ultrafine fibers were prepared [5]. Based on the results of the mentioned research, maximum PEG content in fibers with ΔH_{fus} of 86 KJ/Kg at 58.5°C and ΔH_{cryst} of 65.2 KJ/Kg at 39°C, could reach 70 wt% (Fig. 3).

In the context of fibers manufactured via electrospinning, fibers were constructed by using polyvinylidene fluoride (PVDF) and PEG 1000 mixtures with various PEG 1000 contents. According to the obtained results, the diameter of the electrospun fibers is in direct relation with the PEG 1000 content of the compound. Some of the obtained results are as: 1-PVDF/PEG 1000 mats containing silica, represented good mechanical strength and high values of thermal capacities between 59.2 KJ/Kg and 72.2 KJ/Kg at 38.5°C. A variety of mentioned mats could have various applications in energy storage and the manufacture of thermoregulating fibers [6]. LA/PET (1:1, w/w) ultrafine fibers manufactured via electrospinning method. Fibers had average diameter of 710 nm and they were cylindrical in shape and had smooth surface. Fibers phase change temperatures were 45.1°C and 38.6°C and the corresponding enthalpy changes was 70.8 KJ/Kg and 62.1

KJ/Kg, respectively [7], [8]. Ultrafine fibers were electrospun through stearyl stearate (SS) and polyethylene terephthalate (PET). According to the obtained results, thermal capacities of SS/PET fibers with 10/100-50/100 at 50 °C, increased from 14.3 to 53.8 KJ/Kg [9]. Ultrafine fibers consisted of PET, LA and silica nanoparticles manufactured via electrospinning [10]. In another study, a series of diacid dioctadecyl esters (DADOE) had been synthesized, and subsequently, DADOE fibers with PET were obtained. From the obtained results it was concluded that when the DADOE content of DADOE/PET fibers was less than 50%, the obtained fibers had smooth surface, good morphology and average diameter of fiber increased from 375 nm (pure PET fiber) to more than 1 µm as a result of DADOE content increase. Subsequent heating cycles, DADOE/PET fibers had the thermal stability and capacity of 65 KJ/Kg at 45-48°C [11].

B. Addition of MicroPCMs to Fibers, Textiles and Foams

New and novel methods had been developed in order to add and embed microPCMs to textiles. Researchers by applying new techniques had been able to solve the problem of adding enough PCM inside the fiber structure and improve thermal properties and simultaneous maintaining mechanical properties of fibers. In performed researches, microcapsules containing PCMs impregnated in a binder, subsequently fibers and textiles coated by binder mixture and by applying microPCMs, two and more layers of polymeric foams as thermal insulation pads for shoe insoles had been investigated [12]-[15]. Various textiles applications had been developed after the mentioned researches such as follows: flexible fit clothing for diving in cold water [16], leather products impregnated by microPCMs [17]-[19], textile layers with formed canals filled by microPCMs [20], firefighter clothing, diving, hard work garment, military coveralls, gloves and special shoes [21], [22] and special textiles products containing thermoelectric circuits printed by microPCMs [23]. Initial investigations regarding the manufacturing of foams containing microencapsulated n-alkanes (13-27 C) was conducted and discussed by researchers. In these studies, 20-60 wt% microPCMs added and mixed to a prepolymer compound before hardening in order to ensure homogeneous distribution in the entire system. The manufactured composite foam was considered appropriate for the application as gloves lining, shoes, outdoor wear, car upholstery and medical products [13]. Thermal performance of commercial insulation foams including microPCMs in clothing system had been studied and analytical method obtained for comparison of relative thermal properties of dry garments containing PCM and commercially available dry fabric. Based on the results of this study, it was defined that dry fabrics containing microPCMs could reduce the heat loss of divers during the first phase of diving [24]. In order to investigate the impact of microPCMs on heat and humidity conductivity in textiles, single and double layer textile laminated by PU foam containing 60 wt% of n-octadecane and n-hexadecane microcapsules was manufactured [25]. Other researchers manufactured polyurethane foam incorporated by 12.6 wt% microencapsulated n-octadecane (Fig. 4). Thermal capacity of the foam was measured 12 KJ/Kg between 28°C and 31°C [26].

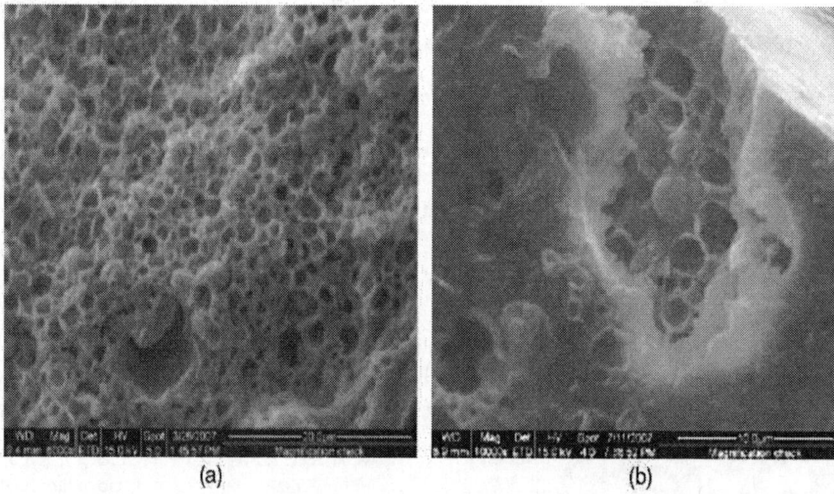

Fig. 4 SEM micrographs of microPCMs/polyurethane composite foam: (a) × 6000 and (b) × 10000 [26]

Soft polyurethane (PU) foams containing 23.6-25.4 wt% n-octadecane, n-nonadecane and n-eicosane microcapsules with styrene-divinylbenzene copolymer shell were manufactured by researchers and the enthalpy of foam containing 25.2% microencapsulated n-octadecane was measured 28 KJ/Kg [27], [28].

C. Direct Addition of PCMs to Textiles, Fibers and Foams

Polyester, cotton, polyamide 66 and wool textiles were

prepared by PEG 600, PEG 1000, 2,2-dimethyl-1,3-propanediol, 2 hydroxymethyl 1-2 methyl 1,3 propanediol aqueous solutions. In comparison to untreated textiles, they had 250% more thermal storage and releasing capacities. Thermal properties of optimized textiles remained stable even after 50 heating and cooling cycles [29]. Also first commercial phase change textile named Neurtratherm® manufactured and introduced by coating with PEG 1000. The phase change property of the manufactured fabric only lasted 20 minutes. Researchers reported that with providing a cross-link on the textiles by PEGs (molecular weight 600-1000) via a reaction by dimethylol dihydroxyethyleneurea under conventional pad-dry-cure conditions, stability of the thermoregulating property could be enhanced. Also, according to the announced results, the manufactured textiles retained their thermoregulating property even after laundering [30]. Fibrous layers (cellulose, polyolefin and fibers mixture) with different structure were studied by PEGs solution (molecular weight = 1000, 1450) containing DMDHEU and a catalyst acid mixture. The researchers found that the textiles thermal properties in proportion to the degree of cross-linking in amorphous and crystalline areas of semi-crystalline PEGs attached to the fibrous layers, is different [31]. Cotton, cotton/polyester and wool textiles designed with heating and cooling effects via a coating on their surfaces by PEG 1000 and PEG 1450 and PEG 600-PEG1000 were added to cotton/polyester textiles by Knife-over-blanket in an experimental printing machine. The applied coating material was a urethane-based synthetic disperse of breathable foam with fine pores containing 45-48% active material. In order to increase the degree of cross-linking, a water miscible non-ionic stabilizing agent was applied. Stabilization operations of coated samples containing 17% PCM performed in an oven for 15 minutes at 60-80°C. According to obtained results between 1°C and 34°C temperature ranges, textiles containing PEG 600-PEG 1000 stored and released 22 KJ/Kg heat [32] (Fig. 5).

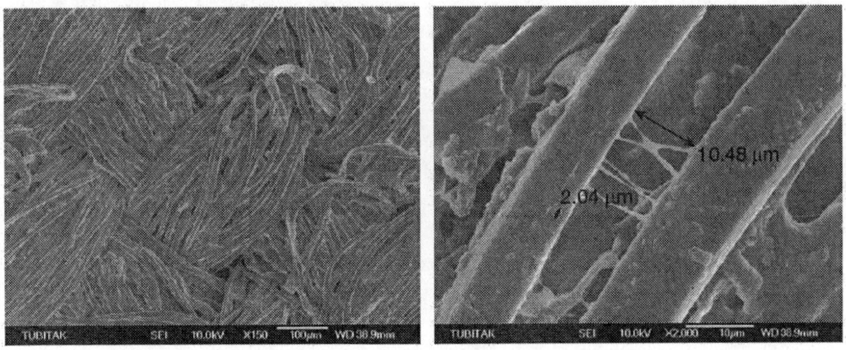

Fig. 5 SEM micrograph of fabric finished by PCM [32]

Polylactic acid (PLA) was invented and in commercial applications of textiles is growing in interest day-by-day [33]. The effect of PEG coating on thermoregulating, antistatic, air permeability, and mechanical properties of PLA textile was investigated. In order to obtain supersaturated PEG 1500 with acceptable stability characteristics by crosslink with DMDHEU, PEG 1500 was fixed on the surface of PLA textile at temperature lower than PLA melting point. After three washing cycles, heat absorbed by PLA textile with 50 wt% PEG was 35 KJ/Kg between 39°C and 56°C [34]. Other researchers widely used PEGs for direct addition to fiber or textile matrix. Although mentioned materials are directly being added to textiles, they are not desirable for applications requiring textile dry cleaning. Studies regarding the application of PEGs for textiles were largely focused on attaching PEGs to textiles, which is related to PEG molecular weight, crosslinking agent and catalyst concentration, curing conditions and fiber polymer type. In addition to obtained thermal characteristics, by application of PEG, textiles resistance against abrasion and pilling also improves [35]. Other researchers became pioneers in the direct addition of PEGs to fibers in order to improve thermoregulating textiles [29], [36]. Composite fibers from PEG with polypropylene (PP), polyethylene terephthalate (PET), or ethylene-vinyl acetate were prepared via physical mixing by controlling the spinning parameters and components [37]. PET copolymerized by various molecular weight PEGs (300, 600, 1000, 2000, 4000 and 6000) and the fibers were melt spun from the copolymer. When the compound included 45 wt% of PEG4000-PEG6000, the enthalpy for fibers measured 26.9 KJ/Kg between 19.6°C and 49.3°C [38]. Thermoregulating fiber (PEG based shape memory thermoplastic polyurethane) obtained from the mentioned experiment at temperatures above PEG melting point still had a certain mechanical strength, because the hydrogen bonded hard, the segment which reacts as a physical crosslink limited the free movement of PEG soft segment. The mentioned fiber had thermal capacity of 100 KJ/Kg at 44.7 °C [39]. Other researchers have added PEG inside wool fiber and investigated the effect of surface energy and charge of wool fiber in order to obtain a high amount of cross-linked PEG with acceptable stability characteristics. These researchers impregnated chlorinated, fluorinated and untreated wool fabrics in 50% PEG 1000 bath. Padded textiles were dried for 5 minutes at 85°C and then cured at 120°C for 2 minutes. The samples were washed by 1% phosphate based detergent at 60°C for 10 minutes and subsequently soaked in water for 5 minutes at 25°C and dried later. The thermal capacity of fluorinated textiles between 31.8°C and 47.1°C was 18 KJ/Kg and for chlorinated textiles, it was 10 KJ/Kg between 30.8°C and 43.2°C [40]. Multicomponent fiber consisted of paraffin waxes such as n-octadecane, n-nonadecane, n-eicosane, n-heneicosane or their mixtures with melting points between 20°C and 50°C had been developed in fiber core by melt spinning. In performed researchers fiber forming polymers were consisted of polyesters, polyamides, polycarbonates, polyoxymethylenes and polyacrylates, because these polymers did not solve at temperatures above PCMs melting or softening points, and therefore, prevent PCM materials leakage. Multicomponent fibers had thermal capacities of 20-40 KJ/Kg between 0°C and 50°C. Based on the obtained results from the mentioned experiments, multicomponent thermoregulating fibers could be used by other types of fibers such as synthetic fibers, cotton, wool or viscose [41]. Additional research in this field was continued by other researchers and based on performed researches, it was found that HDPE with high molecular weight was a suitable viscosity modifier for paraffin waxes and fibers containing 15-42 wt% n-octadecane had thermal capacities of 26 KJ/Kg and 86 KJ/Kg and tenacity of 33 CN/Tex to 16 CN/Tex [42].

III. Conclusion

Customer satisfaction, novel products manufacturing and ensuring a fair income is desired for textiles industry producers and activists, such as other industrial fields, and therefore, manufacturing textiles containing PCMs, textiles which are capable of absorbing, storing and gradual releasing of thermal heat, had been taken under account even more than before. According to the results of the performed researches regarding PCMs and the way of applying them to textiles, electrospinning fibers by use of these materials and the application of microencapsulated PCMs to textiles had priority in comparison to their direct incorporation to textiles and foams. Since PCM materials change phase (solid-liquid, liquid-solid) with thermal fluctuations, therefore in order to save, stabilize and optimal use of them, microencapsulation method inside a flexible polymeric shell and also electrospinning fibers with these materials is recommended. Since the phase change amplitude of the PCMs is very different, so in the textiles industry, the application of PCMs according to body temperature changes will be preferred. The application of PCMs in textiles improves their thermal efficiency and is an effective action in improving the quality of fabrics consumed by humans. In order to provide balance in body temperature, especially in cold environments, it is inevitable to increase the ambient thermal energy consumption or put on more clothes; therefore generally, the application of PCMs in the textile industry causes the related costs to reduce provide economic savings for consumers. By the use of PCMs due to their repeatability of phase change and reduction of human need to supply thermal energy from fossil fuels, it can be stated that these materials have positive effects on the environment.

Acknowledgment

Authors kindly appreciate World Academy of Science, Engineering, and Technology (ICTE 2017) for providing such a valuable platform in order to share new and novel ideas worldwide in the textiles field.

References

[1] G. Erkan, "Enhancing The Thermal Properties of Textiles With Phase Change Materials" in *RJTA, 8*(2), 2004, 57-64.
[2] S. Mondal, "Phase change materials for smart textiles – An overview" in *Applied Thermal Engineering, 28*, 2008, 1536–1550.
[3] N. Bhardwaj, S. C. Kundu, "Electrospinning: a fascinating fiber fabrication technique" in *Biotechnol. Adv., 28*, 2010, 325–347.
[4] J. T. McCann, M. Marquez, Y. N. Xia "Melt coaxial electrospinning: a versatile method for the encapsultion of solid materials and fabrication of phase change nanofibers" in *Nano Lett., 6*, 2006, 2868–2872.
[5] C. Chen, L. Wang, Y. Huang, "Electrospinning of thermo-regulating ultrafine fibers based on polyethylene glycol/cellulose acetate composite" in *Polymer, 48*(18), 2007, 5202–5207.
[6] T. T. T. Nguyen, J. S. Park, "Fabrication of electrospun nonwoven mats of polyvinylidene fluoride/polyethylene glycol/fumed silica for use as energy storage materials" in *J. Appl. Polym. Sci, 121*, 2011, 3596–3603.
[7] C. Chen, L. Wang, Y. Huang, "Morphology and thermal properties of electrospun fatty acids/polyethylene terephthalate composite fibers as novel form-stable phase change materials" in *Sol. Energ. Mater. Sol. C, 92*, 2008, 1382–1387.
[8] C. Chen, L. Wang, Y. Huang, "Electrospinning, a novel shape-stabilized PCM: electrospun ultrafine fibers based on lauric acid/polyethylene terephthalate composite" in *Mater. Lett., 62*, 2008, 3515–3517.
[9] C. Chen, L. Wang, Y. Huang, "Ultrafine electrospun fibers based on stearyl stearate/polyethylene terepthalate composite as form stable phase change materials" in *Chem. Eng. J., 150*, 2009, 269–274.
[10] Y. Cai, H. Ke, J. Dong, Q. Wei, J. Lin, Y. Zhao, L. Song, Y. Hu, F. Huang, W. Gao, H. Fong, "Effects of nano-SiO2 on morphology, thermal energy storage, thermal stability, and combustion properties of electrospun lauric acid/PET ultrafine composite fibers as form-stable phase change materials" in *Appl. Energ., 88*(6), 2011, 2106–2112.
[11] C. Chen, S. Liu, W. Liu, Y. Zhao, Y. Lu, "Synthesis of novel solid–liquid phase change materials and electrospinning of ultrafine phase change fibers" in *Sol. Energ. Mater. Sol. C, 96*, 2012, 202–209.
[12] Y. G. Bryant, D. P. Colvin, "Fabric with reversible enhanced thermal properties" in *US Patent, 5,366,807*, 1994, available from http://patft.uspto.gov/.
[13] Y. G. Bryant, D. P. Colvin, "Moldable foam insole with reversible enhanced thermal storage properties" in US Patent, *5,499,460*, 1996, available from http://patft.uspto.gov/.
[14] D. P. Colvin, Y. G. Bryant, "Thermally enhanced foam insulation" in *US Patent, 5,637,389*, 1997, available from http://patft.uspto.gov/.
[15] D. P. Colvin, Y. G. Bryant, J. C. Driscoll, J. C. Mulligan, "Thermal insulating coating employing microencapsulated phase change material and method" in *US Patent, 5,804,297*, 1998, available from http://patft.uspto.gov/.
[16] M. L. Nuckols, "Analytical modeling of a diver dry suit enhanced with microencapsulated phase change material" in *Ocean Eng., 26*, 1999, 547-564.
[17] R. Painter, "Impregnation of leather with micro-encapsulated material" in *PCT patent application, WO0065100A1*, 2000.
[18] D. K. Robinson, J. J. Erickson, M. Redwood, "Leather impregnated with temperature stabilizing material and method for producing such leather" in *US Patent, 6,179,879*, 2001, available from http://patft.uspto.gov/.
[19] I. Renzi, C. Carfagna, P. Persico, "Thermoregulated natural leather using phase change materials: an example of bioinspiration" *Appl. Therm. Eng., 30*, 2010, 1369-1376.
[20] V. Colvin, D. Colvin, "Microclimate temperature regulating pad and products made there from" in *US Patent*, 6,298,907, 2001, available from http://patft.uspto.gov/.
[21] J. K. Choi, J. G. Lee, J. H. Kim, H. S. Yang, "Preparation of microcapsules containing phase change materials as heat transfer media

by in-situ polymerization" in *J. Ind. Eng. Chem., 7*, 2001, 358-362.

[22] W. Lee, "Microcapsule containing phase change material and article having same." in *PCT patent application, WO02053370A1*, 2002.

[23] M. Rock, V. Sharma, "Heating/warming textile articles with phase change components" in *US Patent, 20020086204A1*, 2002, available from http://patft.uspto.gov/.

[24] M.L. Nuckols, "Analytical modeling of a diver dry suit enhanced with microencapsulated phase change material" in Ocean Eng., *26*, 1999, 547-564.

[25] H. Shim, E. A. McCullough, B. W. Jones, "Using phase change materials in clothing" in Text. Res. J., *71*(6), 2001, 495-502.

[26] M. You, X. X. Zhang, W. Li, X. C. Wang, "Effects of MicroPCMs on the fabrication of MicroPCMs/polyurethane composite foams" in Thermochim. Acta, *472*, 2008, 20-24.

[27] M. You, X. X. Zhang, J. P. Wang, X. C. Wang, "Polyurethane foam containing microencapsulated phase–change materials with styrene–divinylbenzene co–polymer shells" in *J Mater Sci,44*, 2009, 3141-3147.

[28] M. You, X. X. Zhang, X. C. Wang, W. Li, W. Wen, "Effects of type and contents of microencapsuled n-alkanes on properties of soft polyurethane foams" in *Thermochim. Acta, 500*, 2010, 69-75.

[29] T. L. Vigo, C. M. Frost, "Temperature adaptable fabrics" in *Text. Res. J., 55*, 1985, 737-743.

[30] T. L. Vigo, J. S. Bruno, "Temperature adaptable textiles" in *Text. Res. J., 57*, 1987, 427-429.

[31] T. L. Vigo, J. S. Bruno, "Thermal analysis of fibrous substrates containing crosslinked polyols" in *Thermochim. Acta, 161* (2), 1990, 339-351.

[32] W. R. Goynes, T. L. Vigo, J. S. Burno, "Microstructure of fabrics chemically finished for thermal adaptability" in Text. Res. J., *60*, 1990, 277-284.

[33] P. R. Gruber, J. J. Kolstad, C. M. Ryan, R. S. Eichen Conn, E. S. Hall, "Melt-stable lactide polymer nonwoven fabric and process for manufacture thereof" in *US Patent, 6,355,772*, 2002, available from http://patft.uspto.gov/.

[34] A. Khoddami, O. Avinc, F. Ghahremanzadeh, "Improvement in poly(lactic acid) fabric performance via hydrophilic coating" in Prog. Org. Coat., *2*(3), 2011, 299-304.

[35] M. Telkes, E. Raymond, "Storing solar heat in chemicals—a report on the Doverhouse" in *Heat Vent, 46*(11), 1949, 80-86.

[36] T. L. Vigo, C.M. Frost, "Temperature-adaptable hollow fibers containing polyethylene glycols" in *J. Coated Fabrics, 12*, 1983, 243-254.

[37] Y. F. Fan, X. X. Zhang, S. Z. Wu, X. C. Wang, "Thermal stability and permeability of microencapsulated n-octadecane and cyclohexane" in Thermochim. Acta, *429*, 2005, 25-29.

[38] J. Hu, H. Yu, Y. Chen, M. Zhu, "Study on phase-change characteristics of PET–PEG co-polymers" in *J. Macromol. Sci. B, 45*(4), 2006, 615-621.

[39] Q. Meng, J. Hu, "A poly(ethylene glycol)-based smart phase change material" in Sol. Energ. Mater. Sol. C, *92*, 2008, 1260-1268.

[40] F. Ghahremanzadeh, A. Khoddami, C. M. Carr, "Improvement in fastness properties of phase-change material applied on surface modified wool fabrics" in *Fiber Polym.,11*(8), 2010, 1170-1180.

[41] A. Nihlstrand, I. Gabrielii, B. Hagström, "Multi-component fibers" in US Pat. Appl., *20110027568*, 2011, available from http://www.freepatentsonline.com/.

[42] M. T. I. Mollah, "Experimental study on temperature regulating bi-component fibres containing paraffin wax in the core", Report no.: *2009.7.2*, MS Thesis, *The Swedish School of Textiles, University of Boras*, 2009, 23.

Finite Element Modeling of Heat and Moisture Transfer in Porous Material

V. D. Thi, M. Li, M. Khelifa, M. El Ganaoui, Y. Rogaume

Abstract**—**This paper presents a two-dimensional model to study the heat and moisture transfer through porous building materials. Dynamic and static coupled models of heat and moisture transfer in porous material under low temperature are presented and the coupled models together with variable initial and boundary conditions have been considered in an analytical way and using the finite element method. The resulting coupled model is converted to two nonlinear partial differential equations, which is then numerically solved by an implicit iterative scheme. The numerical results of temperature and moisture potential changes are compared with the experimental measurements available in the literature. Predicted results demonstrate validation of the theoretical model and effectiveness of the developed numerical algorithms. It is expected to provide useful information for the porous building material design based on heat and moisture transfer model.

Keywords**—**Finite element method, heat transfer, moisture transfer, porous materials, wood.

NOMENCLATURE

c_m [kg/(kg.°M)]	moisture storage capacity
c_p [J/(kg.K)]	specific heat
D_m[kg/(m.s.°M)]	moisture diffusion coefficient
H [m]	height of specimen
h_C [W/(m^2.K)]	convective heat transfer coefficient
h_m [kg/(m^2.s.°M)]	convective moisture transfer coefficient
h_{LV} [kJ/kg]	heat of phase change
L [m]	thickness of specimen
M [kg/kg]	moisture content
m [°M]	moisture potential
T [K]	temperature
t [s]	time
x [m] and y [m]	location coordinates
x_1 [m]	inner distance
x_2 [m]	outer distance
ε	ratio of vapor diffusion coefficient to coefficient of total moisture diffusion
λ [W/(m.K)]	thermal conductivity
δ [$kg_{moisture}$/(kg.K)]	thermal gradient coefficient
ρ [kg/m^3]	dry solid density
γ [J/kg]	latent heat
in	inner reference
out	outer reference
0	initial
∞	ambient atmosphere

Van Diem Thi, Mengya Li, Mohammed El Ganaoui, and Yann Rogaume are with the University of Lorraine, LERMAB, EA 4370, IUT de Longwy, 54400 Cosnes et Romain, France.

Mourad Khelifa is with the University of Lorraine, LERMAB, EA 4370, 27 rue Philippe Séguin, CS 60036, 88026 Epinal, Cedex, France, (e-mail: mourad.khelifa@univ-lorraine.fr).

I. INTRODUCTION

THE behavior of porous building material such as wood subjected to different environmental conditions is highly dependent on its material structure and chemical composition. The quantity of water inside wood affects its thermo-physical properties and geometric shape. Being porous, hygroscopic and orthotropic material, the heat and moisture transfer in timber makes a complex system of coupled mechanisms. Moreover, if the moisture content changes over time, this may induce distortions, residual stresses, and failures in timber products. In this case, almost all properties in wood as well as the wood–water relations are temperature-dependent. For these reasons, it is of particular interest to have well knowledge of these variables in various timber structures. The main objective of this article is to achieve an analysis of the heat and moisture transfers in porous materials by means of coupled model adopted for finite element analysis.

In the last decades, to improve building energy efficiency and indoor air quality, the coupled heat and moisture transfers in porous materials have been widely studied due to its presence in many industrial applications. Particularly in timber building, researches have been performed [1]-[3]. To evaluate coupled heat and moisture transfer in porous building materials, Abahri et al. proposed a one-dimensional model [1]. For the numerical models, it was found that the comparisons with experimental results give a good correspondence between the distribution of temperature and moisture content inside the porous slab. Chang et al. also described an analytical solution to coupled heat and moisture diffusion transfer in porous materials. For the same application, the obtained results for the wooden slab are similar with published analytical solutions using the decoupling technique [2]. In [3], Younsi et al. present a mathematical model based on a set of coupled, heat and mass transfer equations presented by Luikov [4]. They have also studied the effect of heat and mass transfer parameters through the Biot number on heat and moisture transfer. It was concluded that the governing dimensionless parameters have a significant influence on the kinetics of the heat and moisture transfer. To analyze the heat and moisture transfers in capillary porous media, Luikov [5] proposed a model assuming the same analogy between moisture transport and heat transfer. This model was used by several researchers [6], [7].

To solve the obtained coupled system for temperature and moisture potentials, many authors used both theoretical and numerical approaches. The solution of the governing partial differential equations depends on the considered specific problem. Generally, all approaches are based on the use of transformation functions. Chang et al. [3] have introduced a potential function corresponding to their equations system, using a change of variables for temperature gradient and moisture rate. They used the boundary condition of Neumann type and compared their predicted results with those of Thomas et al. [8]. This paper presents a numerical strategy applied to solve the partial differential equations. It concerns two-dimensional heat and moisture transfer through porous building materials. The finite element solutions were performed using two different boundary conditions. Natural hygrothermal boundary conditions proposed by Chang et al. [2] and non-symmetric Dirichlet boundary conditions developed by Abahri et al. [1] are considered. A numerical approach by means of incremental finite element method will be presented. In order to validate the proposed finite element model, a numerical procedure and a staggered technique to resolve the coupling between temperature and moisture are given in details. Moreover, the comparisons of the obtained numerical results with other models given by Abahri et al. [1], Chang et al. [2] and Younsi et al. [3] were performed.

II. Analytical Formulation

A. Used Assumptions

A typical heat and moisture transfer problem involves a two-dimensional capillary porous specimen (timber) in local coordinates, initially at uniform temperature and uniform moisture content. The boundaries are in contact with the hot surrounding gas, thus resulting in a convection boundary condition for both the temperature and moisture potential as shown in Fig. 1.

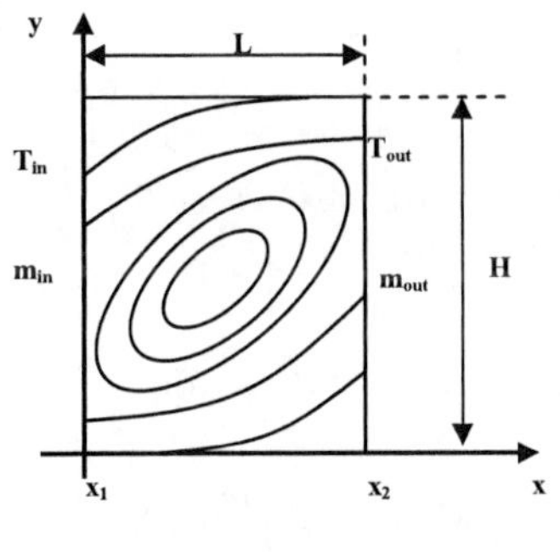

Fig. 1 Coordinate system for porous building material

Heat and moisture transfer in porous media subjected to convective boundary conditions can be modeled using Luikov's approach presented in [5]. The following assumptions are assumed in this work:

- The timber specimen is homogenous and the thermo-physical properties are constant.
- The initial values of moisture and temperature in wood were uniform.
- The shrinkage and the degradation of the timber sample were neglected.

B. Governing Equations

The constitutive equations describing two-dimensional heat and moisture transfer for wood are given by [9] and [10] as:

Heat transfer

$$\rho c_p \frac{\partial T}{\partial t} = \lambda\left(\frac{\partial^2 T}{\partial x^2} + \frac{\partial^2 T}{\partial y^2}\right) + \rho c_m (\varepsilon . h_{LV} + \gamma)\frac{\partial m}{\partial t} \quad (1)$$

Moisture transfer

$$\rho c_m \frac{\partial m}{\partial t} = D_m\left(\frac{\partial^2 m}{\partial x^2} + \frac{\partial^2 m}{\partial y^2}\right) + \delta . D_m\left(\frac{\partial^2 T}{\partial x^2} + \frac{\partial^2 T}{\partial y^2}\right) \quad (2)$$

T[K] is the temperature, m[°M] is the moisture potential, t[s] is time, x[m] and y[m] are the location coordinates, c_p[J/(kg.K)] and c_m [kg/(kg.°M)] are heat and moisture capacities of the medium, λ [W/(m.K)] and D_m [kg/(m.s.°M)] are thermal and moisture diffusion coefficients respectively, ρ [kg/m^3] is the dry solid density, ε is the ratio of vapour diffusion coefficient to coefficient of total moisture diffusion, γ [J/kg] is the latent heat, δ [$kg_{moisture}$/(kg.K)] is the thermal gradient coefficient and h_{LV} [kJ/kg] is the heat of phase change.

C. Initial and Boundary Conditions

The associated hygrothermal boundary and initial conditions are listed below to complete the numerical formulation of the problem:

1. Non-Symmetric Dirichlet Boundary Conditions Adopted in [1]:

a. Initial Conditions

$$T(x,y,t=0)=T_0; \quad m(x,y,t=0)=m_0 \quad (3)$$

$$T(x=0,y,t)=T_{in};\ T(x=L,y,t)=T_{out}$$
$$m(x=0,y,t)=m_{in};\ m(x=L,y,t)=m_{out} \quad (4)$$

The moisture potential m is related to the moisture content M by:

$$M=c_m \times m \quad (5)$$

c_m [kg/(kg.°M)] is the specific mass capacity.

b. Boundary Conditions

$$\frac{\partial T}{\partial y}(x,y=0,t)=0; \quad \frac{\partial T}{\partial y}(x,y=H,t)=0;$$
$$\frac{\partial m}{\partial y}(x,y=0,t)=0; \quad \frac{\partial m}{\partial y}(x,y=H,t)=0 \quad (6)$$

2. Initial and Boundary Conditions Adopted in [2]

To apply the same boundary conditions assumed in [2], we consider an infinitely long rectangular plate, having inner and outer distances x_1[m] and x_2[m], respectively, subjected to symmetrical hydrothermal loads.

a. Initial Conditions

$$T(x,y,t=0)=T_0; \quad m(x,y,t=0)=m_0 \quad (7)$$

$$\begin{aligned} &T(x_1,y,t)=T_\infty; \quad T(x_2,y,t)=T_\infty; \\ &m(x_1,y,t)=m_\infty; \quad m(x_2,y,t)=m_\infty \end{aligned} \tag{8}$$

b. Boundary Conditions

$$\begin{aligned} &\lambda\frac{\partial T}{\partial x}(x_1,y,t)=h_C\left[T(x_1,y,t)-T_\infty\right] \\ &+(1-\varepsilon)h_{LV}h_m\left[m(x_1,y,t)-m_\infty\right] \end{aligned} \tag{9}$$

$$\begin{aligned} &\lambda\frac{\partial T}{\partial x}(x_2,y,t)=-h_C\left[T(x_2,y,t)-T_\infty\right] \\ &-(1-\varepsilon)h_{LV}h_{ml}\left[m(x_2,y,t)-m_\infty\right] \end{aligned} \tag{10}$$

$$D_m\frac{\partial m}{\partial x}(x_1,y,t)+\delta D_m\frac{\partial T}{\partial x}(x_1,y,t)=h_m\left[m(x_1,y,t)-m_\infty\right] \tag{11}$$

$$D_m\frac{\partial m}{\partial x}(x_2,y,t)+\delta D_m\frac{\partial T}{\partial x}(x_2,y,t)=-h_m\left[m(x_2,y,t)-m_\infty\right] \tag{12}$$

$$\begin{aligned} &\frac{\partial T}{\partial y}(x,y=0,t)=0; \quad \frac{\partial T}{\partial y}(x,y=H,t)=0; \\ &\frac{\partial m}{\partial y}(x,y=0,t)=0; \quad \frac{\partial m}{\partial y}(x,y=H,t)=0 \end{aligned} \tag{13}$$

h_C [W/(m^2.K)] is the convective heat transfer coefficient and h_m [kg/(m^2.s.°M)] is the convective moisture transfer coefficient.

Equations (1)-(2) are coupled heat and moisture transfer equations for porous building materials can be solved numerically by the finite element modelling.

III. Finite Element Modeling

A. Numerical Procedure

The Galerkin's weighted residual method is used to discretise the space for (1) and (2). Temperature T is selected as the state variable for heat transfer. The state variable for moisture transfer is chosen as moisture potential, denoted as m. For an element, the state variables are approximated by linearizing three node elements, for example:

$$T(x,y,t)=\sum_{i=1}^{3}N_i(x,y).T_i(t); \quad m(x,y,t)=\sum_{i=1}^{3}N_i(x,y).m_i(t) \tag{14}$$

where $N_i(x,y)$ is the shape functions of triangular finite element, $T_i(t)$ are the time-dependent nodal temperatures and $m_i(t)$ are the time-dependent nodal moistures. Applying Galerkin's weighted residual approach to (1) and (2):

$$\begin{aligned} &\rho c_p\int_\Omega N_i\frac{\partial T}{\partial t}d\Omega=\lambda\int_\Omega N_i\left(\frac{\partial^2 T}{\partial x^2}+\frac{\partial^2 T}{\partial y^2}\right)d\Omega \\ &+\rho c_m\left(\varepsilon.h_{LV}+\gamma\right)\int_\Omega N_i\frac{\partial m}{\partial t}d\Omega \end{aligned} \tag{15}$$

$$\begin{aligned} &\rho c_m\int_\Omega N_i\frac{\partial m}{\partial t}d\Omega=D_m\int_\Omega N_i\left(\frac{\partial^2 m}{\partial x^2}+\frac{\partial^2 m}{\partial y^2}\right)d\Omega \\ &+\delta.D_m\int_\Omega N_i\left(\frac{\partial^2 T}{\partial x^2}+\frac{\partial^2 T}{\partial y^2}\right)d\Omega \end{aligned} \tag{16}$$

Inserting boundary conditions, applying the phase equilibrium relations and rearranging (15) and (16), the resulting system of ordinary differential equations is obtained:

$$\left[C_T\right]\{\dot{T}\}+\left[K_T\right]\{T\}=\left[F_T\right]\{m\} \tag{17}$$

$$\left[C_m\right]\{\dot{m}\}+\left[K_m\right]\{m\}=\left[F_m\right]\{T\} \tag{18}$$

where $[C_T],[K_T],[F_T],[C_m],[K_m]$ and $[F_m]$ are corresponding matrices consisting of coefficients derived from (15) and (16).

B. Numerical Integration

The coupled response of (17) and (18) may be solved using either the dynamic explicit scheme or the implicit scheme. Using an implicit global resolution scheme, the moisture field derived in the previous time step is first used in the temperature equation (17) and a new temperature field is obtained without changing the moisture. The temperatures obtained are then used in (18) to calculate the new moisture potential.

Evaluating the temperature field and moisture potential at state (n+1) from (17) and (18) using the implicit solution approach described above, we obtain:

$$[C_T]\{T^{n+1}\}+\Delta t.[K_T]\{T^{n+1}\}=\Delta t.[F_T]\{m^{n+1}\}+[C_T]\{T^n\} \tag{19}$$

$$[C_m]\{m^{n+1}\}+\Delta t.[K_m]\{m^{n+1}\}=\Delta t.[F_m]\{T^{n+1}\}+[C_m]\{m^n\} \tag{20}$$

$$\{T^{n+1}\}=\left[[C_T]+\Delta t.[K_T]\right]^{-1}\left(\Delta t.[F_T]\{m^{n+1}\}+[C_T]\{T^n\}\right) \tag{21}$$

$$\{T^{n+1}\}=[A_T]^{1}\left(\Delta t.[F_T]\{m^{n+1}\}+[C_T]\{T^n\}\right) \tag{22}$$

Substituting (22) into (20), we obtain:

$$\begin{aligned} &\left([C_m]+\Delta t.[K_m]\right).\{m^{n+1}\}=\Delta t.[F_m][A_T]^{-1}\left(\Delta t.[F_T]\{m^{n+1}\}+[C_T]\{T^n\}\right) \\ &+[C_m]\{m^n\} \end{aligned} \tag{23}$$

$$\begin{aligned} &\left([C_m]+\Delta t.[K_m]-\Delta t^2.[F_m][A_T]^{1}[F_T]\right).\{m^{n+1}\}= \\ &\Delta t.[F_m][A_T]^{1}[C_T]\{T^n\}+[C_m]\{m^n\} \end{aligned} \tag{24}$$

$$\{m^{n+1}\}=[A_m]^{1}\{B_m\} \tag{25}$$

$$\{T^{n+1}\}=[A_T]^{1}\{B_T\} \tag{26}$$

where:

$$[C_T]=\rho c_p\int_\Omega\{N_i\}<N_j>d\Omega \tag{27}$$

$$[K_T]=\lambda\int_\Omega\left[\left\{\frac{\partial N_i}{\partial x}\right\}<\frac{\partial N_j}{\partial x}>+\left\{\frac{\partial N_i}{\partial y}\right\}<\frac{\partial N_j}{\partial y}>\right]d\Omega \quad (28)$$

$$[F_T]=-D_m(\varepsilon.h_{LV}+\gamma)\int_\Omega\left[\left\{\frac{\partial N_i}{\partial x}\right\}<\frac{\partial N_j}{\partial x}>+\left\{\frac{\partial N_i}{\partial y}\right\}<\frac{\partial N_j}{\partial y}>\right]d\Omega \quad (29)$$

$$[C_m]=\rho c_m\int_\Omega\{N_i\}<N_j>d\Omega \quad (30)$$

$$[K_m]=D_m\int_\Omega\left[\left\{\frac{\partial N_i}{\partial x}\right\}<\frac{\partial N_j}{\partial x}>+\left\{\frac{\partial N_i}{\partial y}\right\}<\frac{\partial N_j}{\partial y}>\right]d\Omega \quad (31)$$

$$[F_m]=-\delta.D_m\int_\Omega\left[\left\{\frac{\partial N_i}{\partial x}\right\}<\frac{\partial N_j}{\partial x}>+\left\{\frac{\partial N_i}{\partial y}\right\}<\frac{\partial N_j}{\partial y}>\right]d\Omega \quad (32)$$

$$[A_T]=[C_T]+\Delta t.[K_T] \quad (33)$$

$$\{B_T\}=\Delta t.[F_T]\{m^{n+1}\}+[C_T]\{T^n\} \quad (34)$$

$$[A_m]=[C_m]+\Delta t.[K_m]-\Delta t^2.[F_m][A_T]^{-1}[F_T] \quad (35)$$

$$\{B_m\}=\Delta t.[F_m][A_T]^{-1}[C_T]\{T^n\}+[C_m]\{m^n\} \quad (36)$$

The known solutions at step (n) are assumed to be independent of iterations. The start vectors are chosen as the values at the last converged state, i.e.

$$\begin{aligned}\{T^{n+1}\}^{(i)}&=\{T^n\}^{(i-1)}+\delta\{T\}^{(i)}\\ \{m^{n+1}\}^{(i)}&=\{m^n\}^{(i-1)}+\delta\{m\}^{(i)}\end{aligned} \quad (37)$$

The iteration scheme is stopped when the convergence criterion is verified and the next step is computed. The convergence criterion is given by the scalar product as follow:

$$\begin{aligned}\left\|\{T^{n+1}\}^{(i)}-\{T^n\}^{(i-1)}\right\|&\le 10^{-10}\\ \left\|\{m^{n+1}\}^{(i)}-\{m^n\}^{(i-1)}\right\|&\le 10^{-10}\end{aligned} \quad (38)$$

The above system of equations (19), (20) is solved numerically by using finite element method and a computer program is written in MATLAB environment.

IV. Validation of the Model

The developed formulation in Section III is used in the finite element method in order to analyze heat and moisture transfer in porous building materials. The results from the calculations will be in the following compared to the experimental tests.

A. Application on a Wood Slab Building Envelope

An application on a wood slab building envelope has been simulated using the material properties given from [1], [2] which are associated to two different boundary conditions described previously in Section II *C* are recapitulated in Table I.

The Finite Element (FE) model of the wood slab of 24×60 mm is modelled by 960 triangular elements consisted of 525 nodes. Figs. 2 and 3 illustrate respectively the temperature and the moisture distribution in wooden slab sample at different times. From Fig. 2, it can be seen that at time t=1000 s, a larger temperature gradient was established on 3/4 width plane of the tested specimen. During a prolonged heat transfer process (Figs. 2 (a)(c)), the temperature distribution evolves rapidly at the beginning and then becomes stable. The temperature difference between the inner region and the outer region decreases more and more. After the heat and moisture transfer in wood slab panel, a moisture gradient was established in response to temperature gradient with maximum moisture potential in the surface region. The moisture content in the outer region decreased from 80 °M moisture potential to around 70 °M at time t = 4000 s (Fig. 3 (c)). We also note that the moisture rate in the centre zone decreased progressively until the moisture potential reaches 70 °M at time t=1000 s (Fig. 3 (b)).

TABLE I
MATERIAL PROPERTIES FOR WOOD

c_m	0.01 kg/(kg.°M)	ε	0.3
c_p	2500 J/(kg.K)	m_0	86 °M
δ	2 °M/K	m_{in}	45 °M
λ	0.65 W/(m.K)	m_{out}	4 °M
ρ	370 kg/m3	h_{LV}	2.5×10^{-9} J/kg
D_m	2.2×10^{-8} kg/(m.s.°M)	γ	0
L	24 mm	T_∞	110 °C
H	60 mm	m_∞	4 °M
T_0	10 °C	h_c	2.25 W/(m^2.K)
T_{in}	60 °C	h_m	2.5×10^{-6} kg/(m^2.s.°M)
T_{out}	110 °C		

Fig. 4 shows the temperature distribution of at the center of wood slab (x=12 mm). It can be seen that the temperature distributions of the three layers during the heat and moisture transfer can be divided in three different phases: I) An initial rising phase of temperature, II) a non-linear rising in temperature at the center of wood panel and III) a quasi-stable phase in temperature at the center of the wall. The FE results are similar to the experimental results of Abahri et al. [1]. From a short time of 690 s, the temperature is stationary and equal to 85.2 °C, which is well predicted by the model.

Fig. 5 shows the comparison between the analytical solution and predicted results of the moisture potential versus time at the center of wood slab. It can be seen that the moisture potential distributions can be separated in three phases: I) An initial slightly linear increasing phase, II) a quasi-stable phase and III) a non-linear decreasing phase. There is a very good correlation between the analytical solution given in [1] and the obtained numerical results. During the heat and moisture transfer in the porous slab, the moisture potential continues to decrease with increase in time, until the end.

For the same example, the moisture evolution at the surface and at the middle of the slab is shown in Fig. 6. It can be seen that the results from the natural hygrothermal boundary conditions developed by Chang et al. [2] agree well, even though the boundary conditions are different than used by Abahri et al. [1]. With regard to the moisture content profile of the wood slab at the surface and at the middle presented in Fig. 6, a good accuracy of the present model can then be verified by a very small difference around 1% between the predicted and the results obtained by Chang et al. [2].

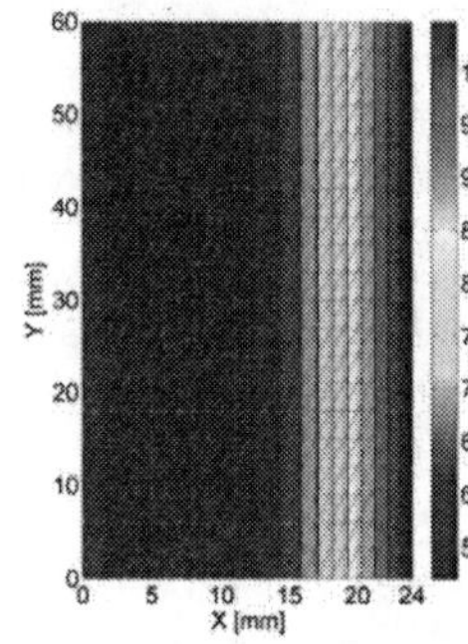

(a) Time t = 100 s

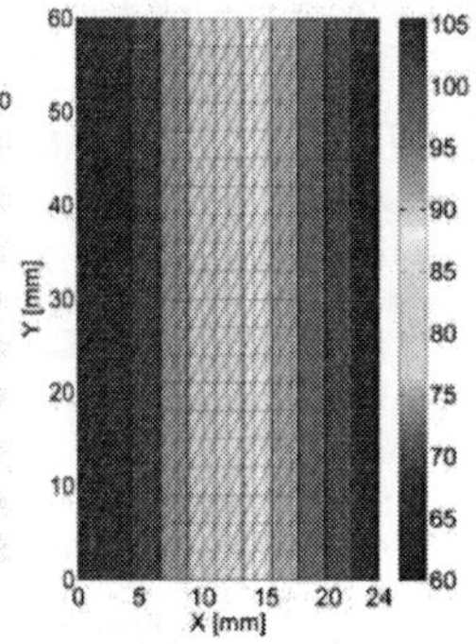

(b) Time t = 500 s

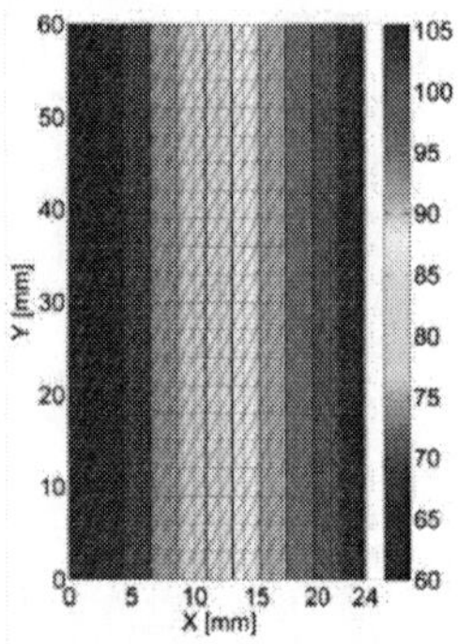

(c) Time t = 1000 s

Fig. 2 Contour maps of temperature distribution at different time

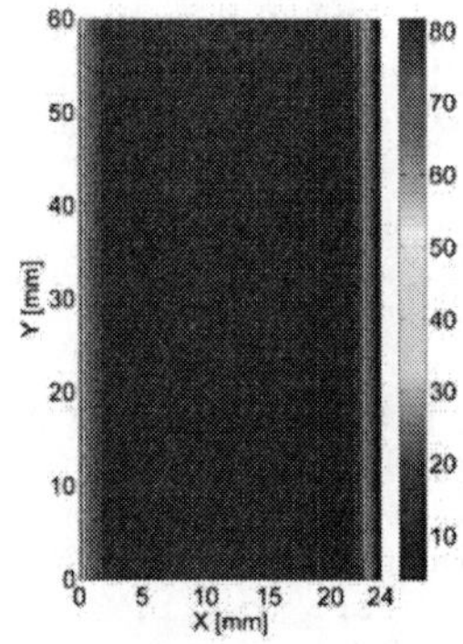

(a) Time t = 100 s

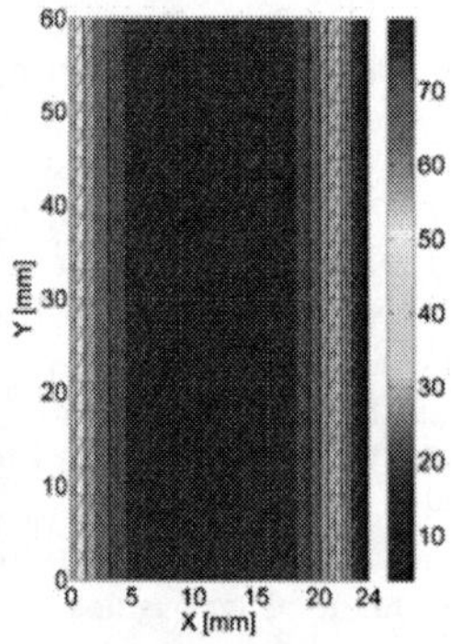

(b) Time t = 1000 s

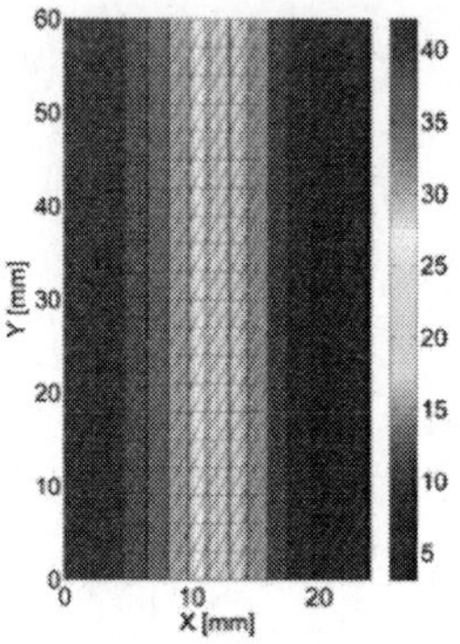

(c) Time t = 4000 s

Fig. 3 Contour maps of moisture potential distribution at different time

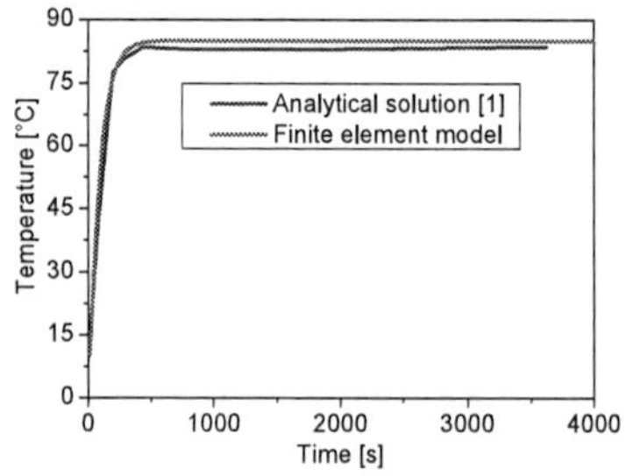

Fig. 4 Curves of temperature versus time at the centre of wood slab

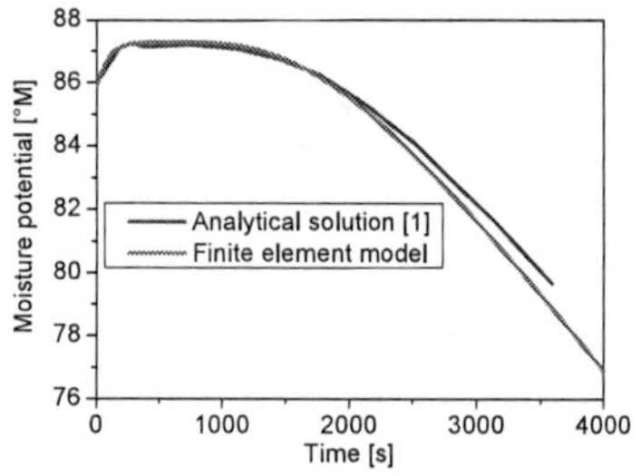

Fig. 5 Curves of moisture potential versus time at the centre of wood slab

The analytical results of the temperature and moisture distributions inside the wood slab are compared with the results obtained by the FE modelling at distinct times, and are illustrated in Figs. 7 and 8, respectively. The transport of capillary water is important in the neighborhood of the inner surface corresponding to the left side of the wood slab. The temperature profile at the center of the panel increases along the time compared to its initial value. This is not the case for the moisture potential value which decreases with time. The maximum differences between the analytical value of the temperature (or moisture potential) and the predicted temperature (or moisture potential) are lower to 0.2%. It can be seen that the FE model shows a good agreement with the analytical model.

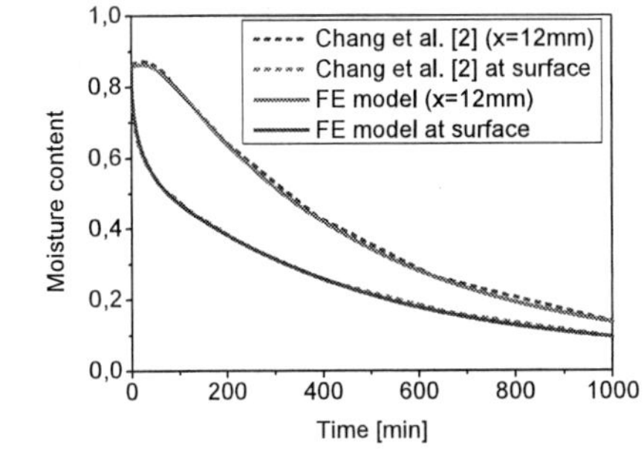

Fig. 6 Curves of moisture content versus time at two different locations

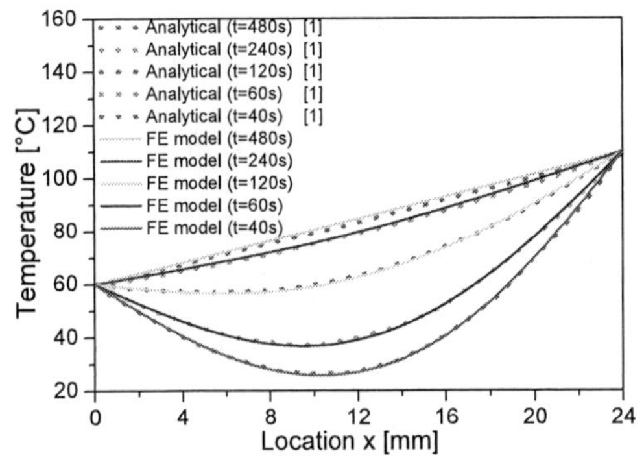

Fig. 7 Curves of temperature versus time at the centre of wood slab

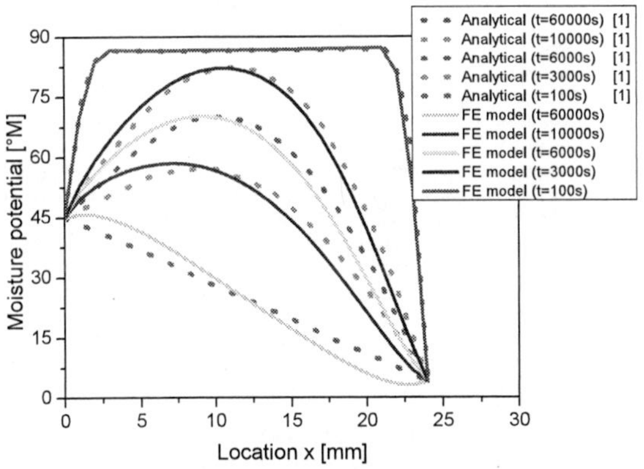

Fig. 8 Curves of moisture potential versus time at the centre of wood slab

Due to the dynamical changing of the meteorological parameters, the heat and moisture transfer in the wall building of 0.05 m thickness changes significantly over the period of a day. To study this impact, the meteorological data for one year period of La Rochelle [1] are taken into account in the simulations. The outdoor conditions (solar radiation, outdoor air temperature and outdoor relative humidity) are taken from meteorological data given in [1]. The initial boundary conditions of temperature and moisture content are 10 °C and 50 °M, respectively. The obtained results are given in Fig. 9. During the heating season, as shown in Fig. 9, the fluctuations of the measured temperatures and the moisture contents are slightly smaller than the calculated values. The simulated results also indicate that the coupling term in energy balance equation has no effect on the temperature values because the effect of the involved moisture potential of the porous material which appears in the coupling term of (1) is not significant. The similar results have been found in [1].

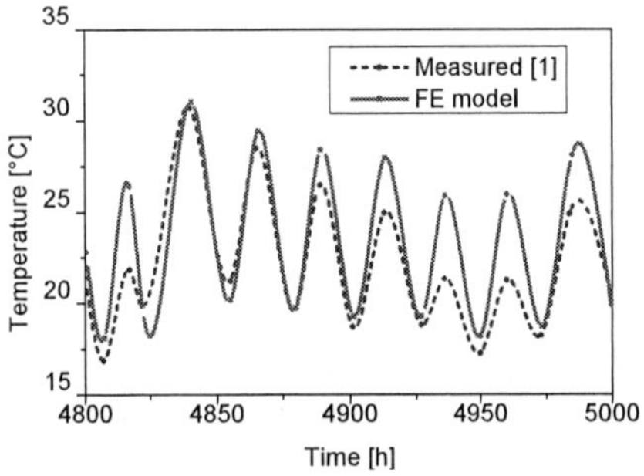

(a) Temperatures

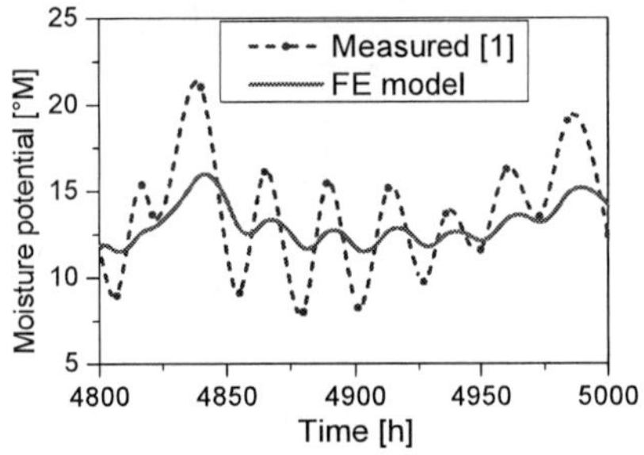

(b) Moisture potentials

Fig. 9 Calculated against measured indoor (a) temperatures and (b) moisture potentials for the porous wall.

B. Simulation of Heat and Moisture Transfer in Wood Material during Thermal Treatment

The geometrical characteristics of the wood specimen are 48 mm width and 49 mm long. The FE mesh consists of 960 three-node triangular elements. The initial and boundary conditions are recapitulated in Table II to complete the analytical and numerical formulations of this example. The same example was analyzed by Chang and Petersen [11] and Hussain and Dincer [12]. The predicted center temperature and moisture distributions inside the spruce wood panel are compared with experimental data given in [11] and the numerical results presented in [3], [12] and are shown in Fig. 10. A good agreement is found between the predicted results and the experimental data for both models (Luikov model [3] and Diffusion model [12]). The absolute error between numerical results and experimental data is less than 5% for temperature value and less than 3% for moisture content value. Therefore, we can state that the proposed FE model is

International Science Index, Materials and Metallurgical Engineering Vol:11, No:4, 2017 waset.org/Publication/10007060 able to estimate the temperature and moisture content distributions inside the wood samples during thermal treatment.

TABLE II
MATERIAL PROPERTIES FOR SPRUCE WOOD SPECIMEN [3]

c_m	0.01 kg/(kg.°M)	m_0	87 °M
c_p	4201.4 J/(kg.K)	ε	0.3
δ	2 °M/K	h_{LV}	2.5×10^{-9} J/kg
λ	0.577 W/(m.K)	γ	0
ρ	856 kg/m3	T_∞	81 °C
D_m	2.2×10^{-8} kg/(m.s.°M)	m_∞	12 °M
L	49 mm	h_c	25 W/(m^2.K)
H	48 mm	h_m	10^{-4} kg/(m^2.s.°M)
T_0	30 °C		

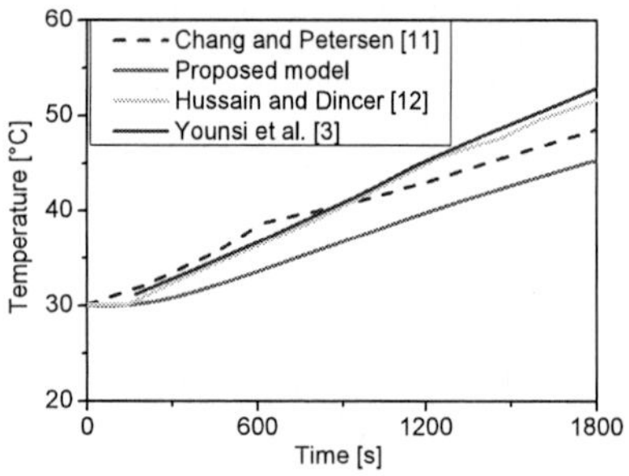

(a) Temperatures

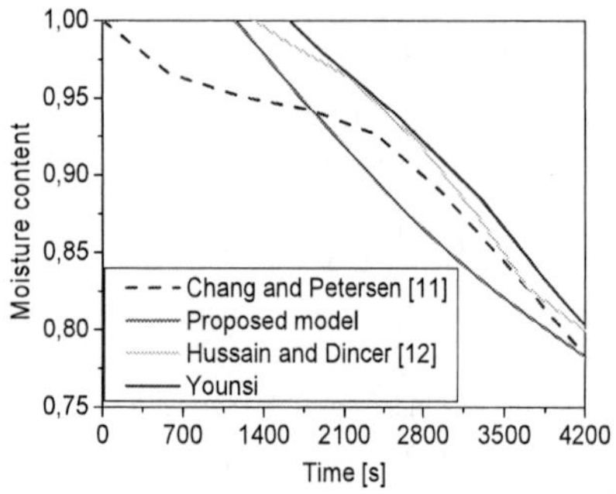

(b) Moisture potentials

Fig. 10 Temperature and moisture content at center of spruce wood sample

V. CONCLUSIONS

The objective of this work is to present a numerical methodology to simulate 2D heat and moisture transfer in porous building materials. In the first step, a generalized mathematical model accounting the governing equations describing the coupled heat and moisture transfers for porous material has been formulated in detail. The associated numerical aspects in the framework of the classical Galerkin-based FE formulation are also discussed. The implicit global resolution scheme as well as the local integration scheme, to compute the values of the temperature and the moisture content at each integration point of each element, is presented.

Several FE simulations using two different boundary conditions are conducted to compare the predicted results with the analytical and experimental measurements available in literature. The quality of the solutions obtained in terms of the hygrothermal fields' distribution, through the porous panel thickness, seems very satisfactory.

Further investigations are now under progress to extend this numerical procedure to 3D case with the hope of simulating the heat and moisture transfers in the porous building panels made by different materials and several layers under complex and severe hygrothermal conditions.

ACKNOWLEDGMENT

The authors gratefully acknowledge the financial support of the French Ministry of High Education and Research, as well as the financial support of the French National Research Agency (ANR) as part of the "Investissements d'Avenir" programme (supported by the Lab of Excellence ARBRE).

REFERENCES

[1] Abahri K, Belarbi R, Trabelsi A. Contribution to analytical and numerical study of combined heat and moisture transfer in porous building materials. Build Environ 2011;46:1354-1360.
[2] Chang W, Weng C. An analytical solution to coupled heat and moisture diffusion transfer in porous materials. Int J Heat Mass Tran 2000;43:3621-3632.
[3] Younsi R, Kocaefe D, Kocaefe Y. Three-dimensional simulation of heat and moisture transfer in wood. Appl Therm Eng 2006;26:1274-1285.
[4] Luikov AV. Systems of differential equations of heat and mass transfer in capillary-porous bodies. Int J Heat Mass Tran 1975; 8:1–14.
[5] Luikov AV. Heat and mass transfer in capillary-porous bodies. Oxford: Pergamon Press; 1966. chap. 6.
[6] Thomas HR, Morgan K, Lewis RW. Afully monolinear analysis of heat and mass transfer problems in porous media. Int J Numer Meth Eng 1980;15:1381-93.
[7] Ribeiro JW. Complete and satisfactory solutions of Luikov equations of heat and moisture transport in a spherical capillary-porous body. Int Commun Heat Mass 2000;27:975-84.
[8] Thomas HR, Lewis RW, Morgan K. An application of the finite element method to the drying of timber. Wood Fiber 1980;11(4):237-43.
[9] Liu JY, Cheng S. Solution of Luikov equations of heat and mass transfer in capillary porous bodies. Int J Heat Mass Tran 1991;34:1747-54.
[10] Qin M, Belarbi R. Development of an analytical method for simultaneous heat and moisture transfer in building materials utilising transfer function method. J Mater Civil Eng ASCE 2005;17(5):492-7.
[11] Chiang W.C., Petersen J.N. Experimental measurement of temperature and moisture profiles during apple drying. Dry Technol 1987;5(1):25–49.
[12] Hussain M.M., Dincer I. Numerical simulation of two-dimensional heat and moisture transfer during drying of a rectangular object. Numerical Heat Transfer A 2003; 43:867–878.

Towards a Complete Automation Feature Recognition System for Sheet Metal Manufacturing

Bahaa Eltahawy, Mikko Ylihärsilä, Reino Virrankoski, Esko Petäjä

***Abstract*—**Sheet metal processing is automated, but the step from product models to the production machine control still requires human intervention. This may cause time consuming bottlenecks in the production process and increase the risk of human errors. In this paper we present a system, which automatically recognizes features from the CAD-model of the sheet metal product. By using these features, the system produces a complete model of the particular sheet metal product. Then the model is used as an input for the sheet metal processing machine. Currently the system is implemented, capable to recognize more than 11 of the most common sheet metal structural features, and the procedure is fully automated. This provides remarkable savings in the production time, and protects against the human errors. This paper presents the developed system architecture, applied algorithms and system software implementation and testing.

***Keywords*—**Feature recognition, automation, sheet metal manufacturing, CAM, CAD

I. Introduction

SHEET metal products vary with their design complexity from very simple ones as with simple cuts or holes, to complicated ones which include nested features (feature means topography or structure in this context) that host other features in other directions [1]. This in turn puts a stress to the design phase. Previous research showed that the design phase can consume up to 70% [2] of the total production cycle. Moreover, at least 50 to 60% of the time consumed by design is actually lost with other tasks related to the design and handling processes rather than the design work itself. It is believed that the design phase and any required modifications are the most costly ones of the production cycle [3], and thus many assisting tools have been suggested to fix for the lost time and effort within this phase.

Computer-aided Drafting (CAD) and Computer-aided Manufacturing (CAM) are typical examples of these tools [4]. CAD software products are used to optimize design and to reduce the design phase time. Many tools exist under this category to suit companies of different sizes, for instance AutoCAD, TurboCAD, CATIA, Solidworks, Inventor, etcetera [5] [6]. Some CAD software would work with different 2d designs, and then they get combined and converted to 3d models during the manufacturing phase, while other products have the ability to work with 3d from the beginning. Minor differences exist between the different suites, and they all are known as being processor, memory and graphics intensive applications that require high specifications. Still in real situation there can be a significant time lost due to lagging, processing, converting, and related file tasks. The other category is CAM which is used to assist the manufacturing process and its related tasks, e.g. planning, storage, tooling, etc. Normally, the work flow starts with CAD, then CAM, and ends with implementation, which compared to previous tasks takes the least time.

Many issues arise within the design phase and need to be considered. Firstly, it is assumed by default that designers know the properties of the products they are working with, sheet metal specifications, and machinery limitation. Another important task for designers to do is to document the design and its features to be ready for manufacturing. These features include but are not limited to thickness, diameter, direction, position, and angle. However, for the bending factor, also known as k-factor, it is usually not known at the design phase since it depends of the bending tools that will be used to for bending. Documentation is one of the most time-consuming processes, because of the amount of details it may include. Moreover, any further modifications that are required due to any limitation will turn the process back to the design phase to start over, and then to documentation after. A very important factor to pay attention to as well is the individual experience and the design's dependability on the designer's style himself. Designers know more details than they document about their own work, and different individuals have different design approaches. This in turn may cause problems with complicated designs if they would be handled by others. Also any missed details in documentation may cause manufacturing errors and render parts waste material. All these reasons indicate the need of sophisticated feature recognition. Feature recognition systems would provide the required assistance during the design phase, since they perform documentation and guidance for the control of the sheet metal processing machine, while including the pre-defined specifications. Many feature recognition systems already exist in different companies that work in the sheet metal industry, trying to simplify the design phase to its basics to speed up the design process. Human intervention within the recognition phase still cannot be avoided, and machines are unable to perform all feature-based tasks completely. Moreover, any shortage or required modification during the final stages would return designs for further processing. In this work with Prima Power we are completing the feature recognition shortages in the existing system. We target to increase the automation level, to

Bahaa Eltahawy and Reino Virrankoski are with the University of Vaasa, Department of Computer Science, P.O. Box 700, FI-65101, Vaasa, Finland (e-mail: {bahaa.eltahawy, reino.virrankoski}@uva.fi).

Mikko Ylihärsilä and Esko Petäjä are with the Prima Power Oy, Metallitie 4, FI-62200, Kauhava, Finland (e-mail: {mikko.yliharsila, esko.petaja}@primapower.com).

reduce the human error, processing and tooling time. One key target is to improve the production process so that the separation between design and manufacturing phases becomes clear. We are doing so by introducing a system that can recognize all typical features with all possible variations and factors for the various CAD-models, thus the manufacturing phase goes totally independent of the design phase. In turn this will save the time required for documentation, tooling and other models' handling tasks. This speeds up the design phase by at least 50% and creates a better independent feature recognition system.

In this paper we introduce our solution for sheet metal feature recognition from the model. Hereby it includes the system, the applied feature types, challenges and limitations we faced during our work, and the ways how we solved them. We introduce the main feature recognition rules and implementation schemes, and finally we provide some study cases with different levels of complexity.

II. The System

The system consists of following entities: CAD system, feature recognition system, and design and testing library. Due to its robustness and flexibility to provide a wide range of functions to utilize, our CAD solution is Siemens PLM NX® (https://www.plm.automation.siemens.com/) modeler [7] [8]. This suite allows us to perform the CAD and CAM functions, and it also provides tooling, handling, programming, processing, cutting, and many other properties that are beneficial to our implementation. The other part of the system is Microsoft Visual Studio® which is an Integrated Development Environment (IDE) (https://www.visualstudio.com/) that we use to program our recognition system, i.e. Feature Recognition Engine©® (FRE) [9]. The third part of the system is the intensive library that we have created to study all considered features, the testing library has been created using SolidWorks® (http://www.solidworks.com/). Finally, the system output is the implementation sheet with the found features, their specifications and the required tooling, thus making the imported design ready for the manufacturing stage.

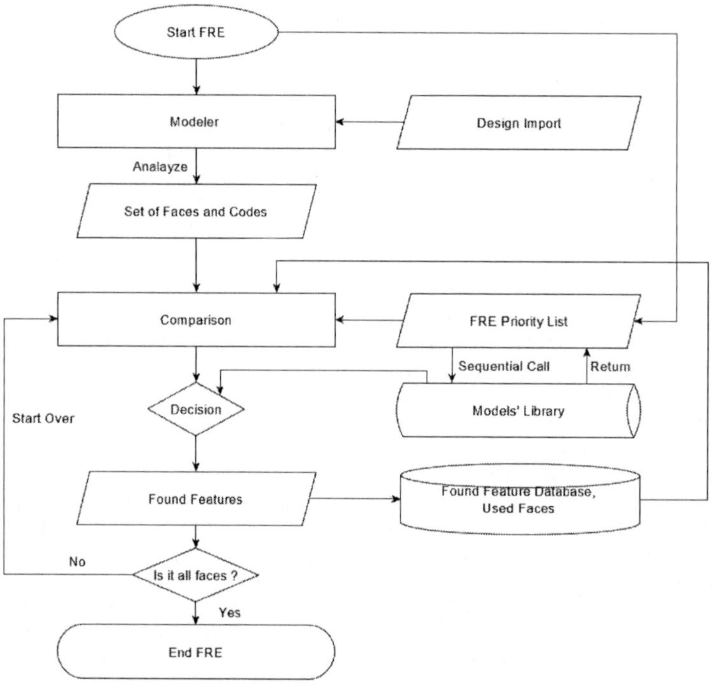

Fig. 1 The flowchart of the developed Feature Recognition System (FRE)

Feature recognition flowchart is presented in Fig. 1. Here FRE system upon importing a certain CAD-design launches the modeler system, to visualize the design and to open it in memory. FRE system will analyze the CAD-design and convert it to a set of faces with a set of codes. Recognition is done by performing a thorough comparison with the pre-defined feature models that are stored in the model library. To avoid conflicts, certain features are given higher priority than others, so that features do not get recognized mistakenly as the most common ones. For instance, louver feature without prioritization will be recognized as a set of bent features. For that reason testing will be done for features with the higher priority at first. FRE therefore starts by knowing which feature it should test at first by consulting the priority list, and then after it performs a call to its model library. Here, FRE will perform all the required comparisons. Matching is done between the input design's found features and the feature models stored in the model library. Once matching is successful, extra tests for validation are performed to ensure a correct identification of a feature. If a feature is found, FRE then will save all the found faces of the found feature with their characteristics in the Found Feature Database (FFD). If matching is unsuccessful, FRE brings the next feature to study from the priority list, then repeats the process to study if such feature exists within the model upon study or not. The process is repeated as many times as needed to recognize all features in the processed model.

III. Types of Features

In the context of sheet metal processing, the word "feature" means topography or structure. As presented in Fig. 2, features [10], [11], [3] exist due to the different methods using within manufacturing, for example with cut, draw, bent, and out of stretching. Many features would exist due to such variations, but in our work we rather are focusing on features that take an offset shape out of the sheet metal plate. Namely, we concern holes, extruded holes, cuts, flanged holes, louvers, lanced bridges, dimples, embosses, all forms of beads, and all forms of bends, and finally nested features [12]. The following 11 features are illustrated in Fig. 2.

1. Hole: It is merely a round hole within the sheet metal. A hole can take different shapes depending on the usage purpose; it can be blank hole as in Fig. 2 A, with fillets as in Fig. 2 C, with chamfers as in Fig. 2 B, or even conical or fully rounded fillet as in Fig. 2 B.
2. Extruded hole: This one is an offset from the metal surface, it normally is of 0.5 cm height, but this can change according to the purpose. Extruded holes share the same variations with simple holes; they can have fillets or chamfers, also they can be cylindrical as in Fig. 2 F or conical as in Fig. 2 E.
3. Cut: They are holes that can take any shape depending on the design. For example they can take rounded rectangular form as in Figs. 2 G-H. Both holes and cuts are the same concept, but what makes difference between them is the ratio and dimensions that form the cut itself.
4. Flanged holes: Similarly as extruded holes and cuts. A flanged hole is an offset from the metal surface that can take any other shape rather than the cylindrical one, as in Fig. 2 I.
5. Louvers: A louver is a window that is protected from 3 sides and open from the forth to allow for air passage, it is manufactured as a punch feature. A louver takes normally the shape presented in Fig. 2 J shape, but it can also take the spherical shape as in Fig. 2 K.
6. Lanced Bridges: It is another punch feature, where the punching tool will form a bracket-shaped "]" form that is offset of the sheet plate. Normally bridges take the form in Fig. 2 L.
7. Dimples: A dimple is a stretched offset feature made to the sheet plate, taking the shape of counter sink of a hole. Dimples can be rectangular, rounded-rectangular shape, or they can go non-uniform taking random shapes, these are presented in Figs. 2 M-O.
8. Embosses: Emboss is the same idea as dimple, but rather it is spherical, as shown in Fig. 2 P.
9. Beads: A bead is formed by pinching a metal piece placed between dies, thus the plate takes the shape of the given metal piece. Beads can take several forms as in Figs. 2 Q-S. Also in these figures beads are straight-shaped form, while they also can take a circular form.
10. Bends: A simple bend is presented in Figs. 2 T, V-X. Bends can also be nested which means a bend following a bend, as in Fig. 2 U.
11. Nested features: They can include features that host other features. Fig. 2 Y is an example of nested features. In this figure, the sheet metal plate hosts a dimple feature, that hosts a lance bridge and louver features as well, moreover the hosted bridge feature itself hosts an extruded hole feature on its upper plane.

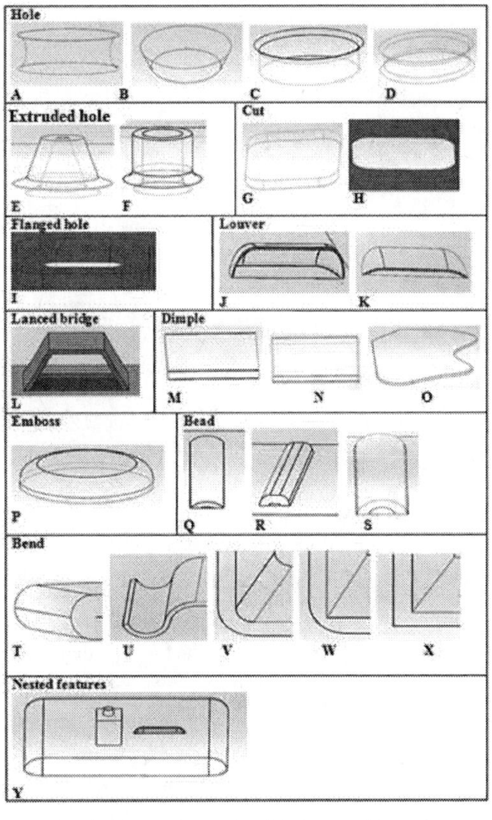

Fig. 2 The set of considered features in this work

IV. Difficulties

The main difficulties in feature recognition are related to designs' inconsistency and common similarities. Firstly, designs have many variations, which need to be considered during modeling. This in turn increases the complexity of the modeling scenario, also the algorithm becomes computationally heavy to execute once the number of combinations increases drastically when all variations must be computed. For example, a very simple hole can be merely a hole, a hole with upper fillet, a hole with lower fillet, a hole with double fillets, a hole with upper chamber, a hole with lower chamfer, a hole with double chamfers, a hole with fillet and chamfer, a conical hole, or a fully rounded fillet-shaped hole. Using coding convention a hole can change its state from 00 to 11, where a digit represents existence of a pattern. Considering that rounding can be either fillet or chamfer gets the result doubled, and by taking into account 2 special cases, a simple hole can take 9 different forms at the end. Similarly, extruded holes can go from 0000 to 1111 with one form, which means $2x2^4$ -1 = 31 different forms at the end if rounding forms are not mixed. With the mixing case however we should consider 3^4 = 81 different variations. For the purpose of complete feature recognition implementation, we have considered all cases that could be available/ applicable. Table I shows the number of variations considering only fillets as rounding pattern. It can be seen from that table that if rounding patterns and other special cases are mixed, then the total number of cases will be at least doubled, i.e. around 1200 recognition cases at least.

The other challenge we have encountered is the similarity that some models share. For example, a basic simple hole can have only one transition, which is a cylinder, a rounded one. A bend as well can only have one rounded curve in a cylindrical form. This in the beginning could cause a lot of confusion, as the system would mix between holes and bends. The same case exactly occurs with other features, for example the louver model in its simplest form has only 3 transitions, in the same time one of the many variations of the bridge model could take a wrong short-cut and come to end-state only after 3 transitions. This will simply show a bridge as louver mistakenly. This can be solved by creating many rules and validators to force the system to go only in a certain direction when searching for a specific feature. Similarities errors are shown in Fig. 3.

TABLE I
NUMBER OF VARIATIONS FOR SELECTED FEATURES

Simple hole	4
Extruded hole	16
Louver	4 + 4 = 8
Lance bridge	256
Dimple	16 x 16 = 256
Bend	4
Bead	16 + (16 x 2) + (16 + 2) = 80 for straight beads, similarly with rounded beads
Emboss	16
Total	636 **simple** cases

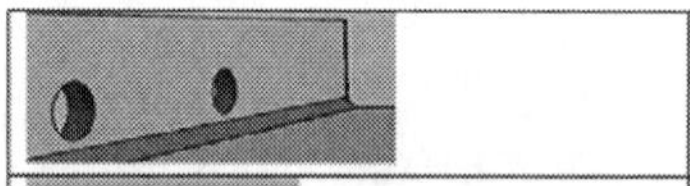	**Bend and Simple hole recognition error**	**Here, the logic for bend and simple hole is: Wall face >>Cylinder >>Wall face**
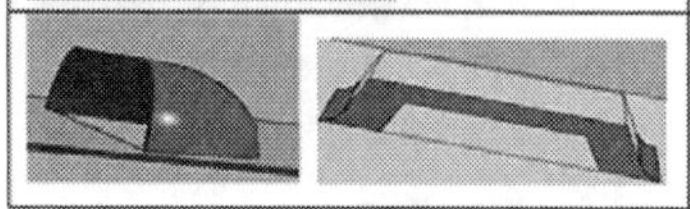	**Louver and Lance bridge recognition error**	**Here, the logic for louver and bridge is: Wall face >>Cylinder >>Planar face >> Cylinder >>Wall face**

Fig. 3 Features with common logical recognition errors

V. FEATURE SET-UP

In order to make successful recognition, the following factors have been considered:

1. The main surface: The recognition algorithm will at first perform a thorough analysis to test all surfaces, to find the surface that will be treated as "the main surface". This surface is found by considering that the surface that has the most number of features hosted within, to be the main surface. This main surface will be connected to the feature nesting as shown hereafter.
2. Surface type: Distinction between surfaces has been done. Surfaces are classified to planar, cylindrical, conical, spherical and torus surfaces. If a surface does not follow the common rule for classification, it is tagged merely as a "surface".
3. Convexity: Many features can get recognized mistakenly if the searching algorithm does not include the convexity of the body it is searching for. Features can exist inwards or outwards, and also the directionality with the main surface will play a factor here. When considering the angle of search and the orientation, we could refine the search to a minor number, so that after the right validators the correct feature will be recognized. The algorithm can search for following convexity types: convex, concave, or when it is not sharp then it will be smooth convex, smooth concave, or when it is beyond certain angles then it is knife convex or knife concave, or when inflecting in the other direction then it searches for inflection.
4. Directionality: For simplicity recognition has been designed to recognize features in only direction, which is upwards. For efficiency the algorithm will run till the end state, then it will perform surface switching, in which the upper face will be considered the bottom one and the bottom face will be considered the upper face. At this stage, the algorithm will perform a thorough scan to find features that could not get recognized in the earlier phase. Finally the output sheet will specify the location of the found feature and its exact direction.
5. Prioritization of features: It is very important to run the algorithm inherently on several stages instead of

performing a one-shot run. Features share many common characteristics, and this might lead to either recognizing a wrong feature, duplication of results, or recognition failure. For instance, bends, louvers, beads, lance bridges or other features may start with cylindrical face in the form of a fillet, and without prioritization the wrong feature will get recognized. As illustrated in Fig. 3, a louver will have a higher priority than a bridge, and also a hole will have a higher priority than a bent, which in fact would have the lowest priority among all other features.

6. Nesting: Features still can exist on other surfaces rather the main one and its opposite direction replica. In other words the features can host other features within one of their faces. After finishing with the main features the algorithm will switch to features individually then treats them as sequence of sub-surfaces to avoid the loss of nested features. Recognition will run on surfaces individually. If other features are found then it will compare them to the main found ones to determine if it is a new feature or just a replica of an already found feature. Nested features will get included within the output, which will include all their characteristics exactly as in the case of main ones.
7. Tolerance: To optimally find all features under study, different tolerance mappings have been considered. It is known from real implementation that different features have different sizes that they exist within. Some features can accept ±1 mm while others can go to 0.001 mm. For that we specified different tolerance mappings according to the feature under study and the tolerance value it can accept. Mathematical explanation is at [13].
8. Forks: To find all faces within a certain feature, the algorithm needs to leave the main route in favor of searching in other directions. For that reason we have included a function called "Fork" that would perform path separation. Normally recognition goes in a certain direction; this will work with features that do not experience similarity from two opposite sides while being in the same face. Other features like louver or lanced bridge require the recognition system to get out of its recognition direction at a certain point to find other surfaces. After finding those side surfaces, the recognition algorithm returns back to its main direction with the found information to be included for successful feature recognition. The return is governed by another function called "Join". Join function combines paths that got separated before, to direct them to the main drawn route again. Unlikely forks suffer from being considered as a minor route compared to the main route, which means that if two features with the same priority would exist then forks will mostly get ignored. As illustrated in Fig. 3, a bridge got recognized as a louver, which is wrong. This was corrected by applying forks in the most minor tracks rather than main ones. For example with the case of louvers and bridges, forks are only considered for fillets, not main bodies.
9. Line and vertex recognition: These are special cases. Typically the recognition algorithm searches for surfaces and bodies. However, in some certain cases this is not applicable. In those cases one needs to search for a line, for example, and then treat it as a body. In some other type of cases there is even no line but a point, vertex, connecting between two surfaces that do not share any common bodies but this single point. Link and vertex recognition is required when dealing with bends and with many of the special cases of dimples.
10. Non-uniform patterns: As mentioned in pact 2, all surface types are uniform, except the ones we explicitly refer as "Surface". The surfaces can take any form and they will not abide with any rules. The problem with surface bodies that the recognition can find one of its faces but not both of them, because as mentioned none of the normal rules can be applied. However, we can tweak this shortage only by considering bodies' thickness. If this is constant despite the body type, the other face will be found. This is shown later when the rules are presented in the next sections.
11. Design constraints: Even though we are targeting to create a system that can deal with all possible cases, few shortages still exist. These include sharpness and angles, and for that recommended ranges are specified during design, so that the recognition system fully functions.

VI. Rules

The recognition algorithm principally works as presented in Fig. 4. The algorithm starts by searching for neighboring faces that have the type upon search, then it checks a set of rules to make sure if this face is the right one or if it is one of many faces sharing the same characteristics. If the result indicates that it is the right face then it will proceed to the next face, and if the result does not indicate that it is the right face then it will not be included for this feature. Faces can include as many validators as needed to distinguish them from other faces. However, not all validators are applicable to all types of faces. Rules that have been used are built by using validators like parallelism, perpendicularity, concentricity, thickness, fully round, or face repeatedly.

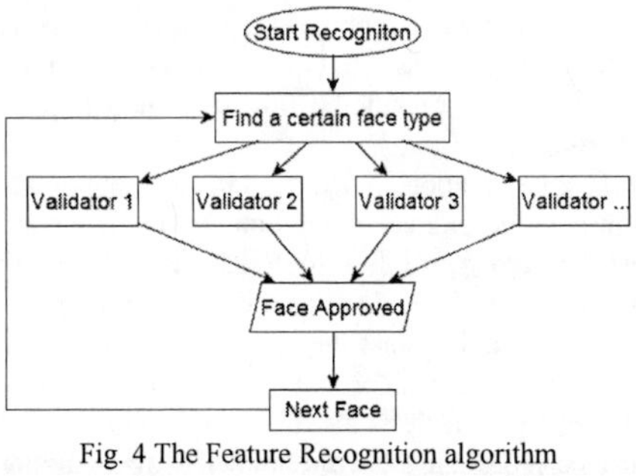

Fig. 4 The Feature Recognition algorithm

1. Parallelism: This will check if surfaces have a parallel relation. For plane it means that the normal (perpendicular) lines from the plane center are rather

parallel. For round shapes like cylinder or cone, parallelism is measured about the concentric lines that form those shapes. For surfaces like chamfers or fillets which are complex shapes because of having two radii, parallelism is measured about the central lines that form the rounding radii lines that directly form that shape, as shown in Fig. 5.

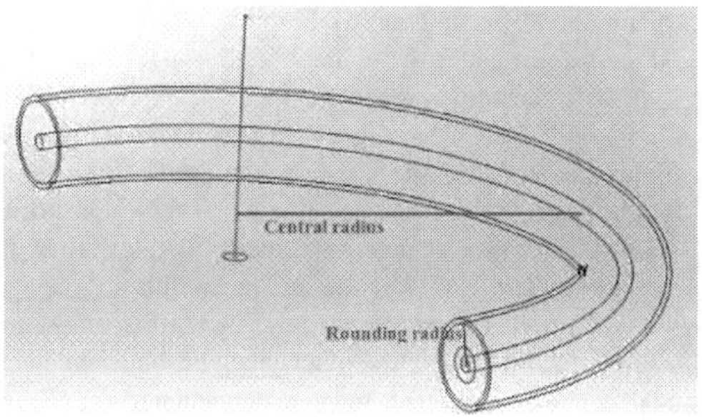

Fig. 5 For tours surfaces, parallelism is measured about the central radius

2. Perpendicularity: This rule is similar to parallelism rule, rather it measures if surfaces have 90 degrees in between. Also perpendicularity unlike parallelism is applicable in different dimensions, so it does not need to be on the same plane.
3. Concentricity: When bodies take one of the rounded forms, this rule does find the adjacent or correspondent faces by examining if they have the same center. By convention surfaces have to be parallel at first.
4. Thickness: Applying parallelism and perpendicularity rules will limit the number of found surfaces. However, when non-uniformity exists, parallelism and perpendicularity rules might be unable to deal with it, and hence we need the rule of thickness. The thickness rule takes a surface as a reference, then runs a test on all other surfaces, till it finds a surface with a match, i.e. a surface with constant thickness from the surface under consideration (reference surface).. This is done by taking into account that the formed body does have a constant thickness across its route. This rule plays a vital role with surfaces tagged as "Surface", which do not have a direction and cannot be included under the normal categories.
5. Surface completion: When dealing with rounded bodies, it is imperative to know if the body is fully rounded or if it just takes a part of a rounded shape. This rule will examine that. The rule is useful to remove the confusion between rounding that any feature can have, e.g. fillets, chamfers, and features that may include a rounding pattern, e.g. holes.
6. Surface repeatedly: For the matter of recognizing bodies that are off-path, the route needs to split into two different routes. Recognition might continue with two or more routes until it reaches the end-state, or it may need to return to the main path again. For the latter case, this rule will test if the face it came across had been examined before i.e. if faces are identical, or the found face had not been examined before, i.e. a new face to examine.

VII. Modeling and Characterization Approach

Feature Modeling aims characterization of all considerable variations a certain feature can take. In Fig. 2 for instance, a simple extruded hole can have 16 possible variations if there are no special cases. To consider these variations, different steps have been taken to characterize the different models and their associated variations. As presented in Fig. 6, and explained below, feature modeling is done as follows:

1. Characterization: In this phase we consider only one feature at a time. Feature characterization has been simplified at first by means of using finite-state machines approach to represent all possible transitions a single feature can take. Details are beyond the scope of this work.
2. Implementation: After characterizing the feature logically in phase 1, phase 2 comes in place where the feature characterization's logic is implemented by means of programming. This will be used in practice later for feature recognition.
3. Identification and Précising: Upon implementing the logic, we need to test it to make sure the feature under study is well recognized. We start with Identification, where we run the program against the feature in its simplest form. For example, if we are modeling Extruded Hole feature, then identification will be done against the model which has no fillets, chamfers, or any special forms. Upon successfulness in identifying the feature in its very basic form, we perform précising by running our program against models with different variations of the same feature. Upon recognition of the feature in its different variations, we proceed to phase 4. Otherwise, we return to phase 1 to modify the characterization logic, and then to phase 2 to modify the implementation.
4. Testing with other features: As mentioned previously, some features have similarities, which result in very close characterization logic. For instance, Louver feature in its simplest form has the same characterization logic as one of the variations of the Bridge feature. In the previous phases, we were considering only the feature under study. In this phase, we consider running the logic and the implementable program of this feature against models hosting other features. The purpose here is to double-check that the characterization logic works only with the feature under consideration, and to avoid any false identification of other features. By now, fully independent feature characterization is achieved.

For the matter of characterizing the different features, the previous four phases have been repeated to model and every feature individually. Now we have set of recognition algorithms and programs, in which each can recognize a certain feature successfully.

5. Nesting: In this phase, we aim identification of the nested structures. This is done by running the program against

the sub-structures of the main structure under study. For instance, a feature can host another feature in one of its faces, so after recognizing the main feature at first, we consider running the recognition algorithm on the feature's different faces themselves to search for any nested features within. To visualize this, in Fig. 2 Y, for instance, the main structure is a model hosting a Dimple feature, in which the Dimple feature's structure hosts Louver and Bridge features as sub-features, in which the Bridge feature's structure as well hosts an Extruded Hole feature as a third-level feature. Here we could recognize all features and the sub-features by running the recognition algorithm over the different structures. That is what we mean by finding nested features. Again, if required, then we may modify the characterization logic at phase 1 if the recognition algorithm would mix between the different features and their hosting ones.

Previous phases have been performed using standard testing models we created earlier for the purpose of testing only. In reality, this is not the case. Thus, further testing has been done with real manufacturing parts extracted from various customers, so that tolerance and flexibility could be adjusted properly to match with real-world parts for effectiveness.

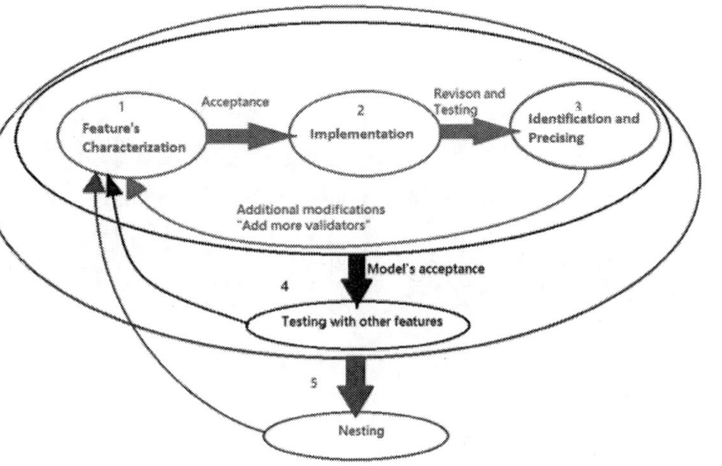

Fig. 6 Feature's modeling and characterization process

VIII. Case Study: A Louver Model

We are here introducing the work that has been done to model louver model. The recognition algorithm is given in Fig. 7, and details are as following:

1. Analyzing surfaces: A lover has mainly 3 bodies with 7 to 9 faces according to the design. As well it can have up to extra 6 faces if it includes fillets, torus or sphere shaped. The surface in between can be a planar or conical surface. If fillets exist, then their main body will take the cylindrical shape, and minor bodies will be either torus or complex surfaces entitled "surface".
2. Analyzing relations between surfaces:
   a. Main cylindrical body: Main outer and inner cylindrical faces are parallel, concentric, perpendicular to the main face, have the same thickness, and are not fully rounded.
   b. Fillets: Main surfaces are parallel, concentric, perpendicular to the main face, have the same thickness, and are not fully rounded. Main surfaces however are not concentric with the main outer and inner cylindrical faces of the main cylindrical body. For minor surfaces, if they are torus-shaped, then they will be parallel, parallel to the main face, concentric and have the same thickness. If sphere-shaped, then they are only concentric and have the same thickness. If surface-shaped, then only the thickness rule is applicable. All fillets are not fully rounded.
   c. Planar faces: They are parallel to each other, and perpendicular to the main face.
3. Analysis of convexity:
   a. To find the main body, the surface is "concave"
   b. To find the main fillet, the surface is "smooth concave".
   c. If a fillet is present, then to find the main body, the convexity rule searches for a surface with convexity type "inflection".
   d. All other bodies are convex, except if there would be a fillet, then there would be inflection.
   e. Reaching the end-state would be either convex or smooth convex if there is a fillet.
4. Technical results: Upon successful recognition, the system could recognize louver feature in a sheet of multiple features. It also gave details to its parameters, including exact position from the center, length, width, height, and thickness. Different bodies were tagged with their radius, face identity number, bending factor, corner rule, and the material for implementation. For the existing curvatures, the system indicates the type of convexity with the type of curve, e.g. line, circle, or non-uniform shape.

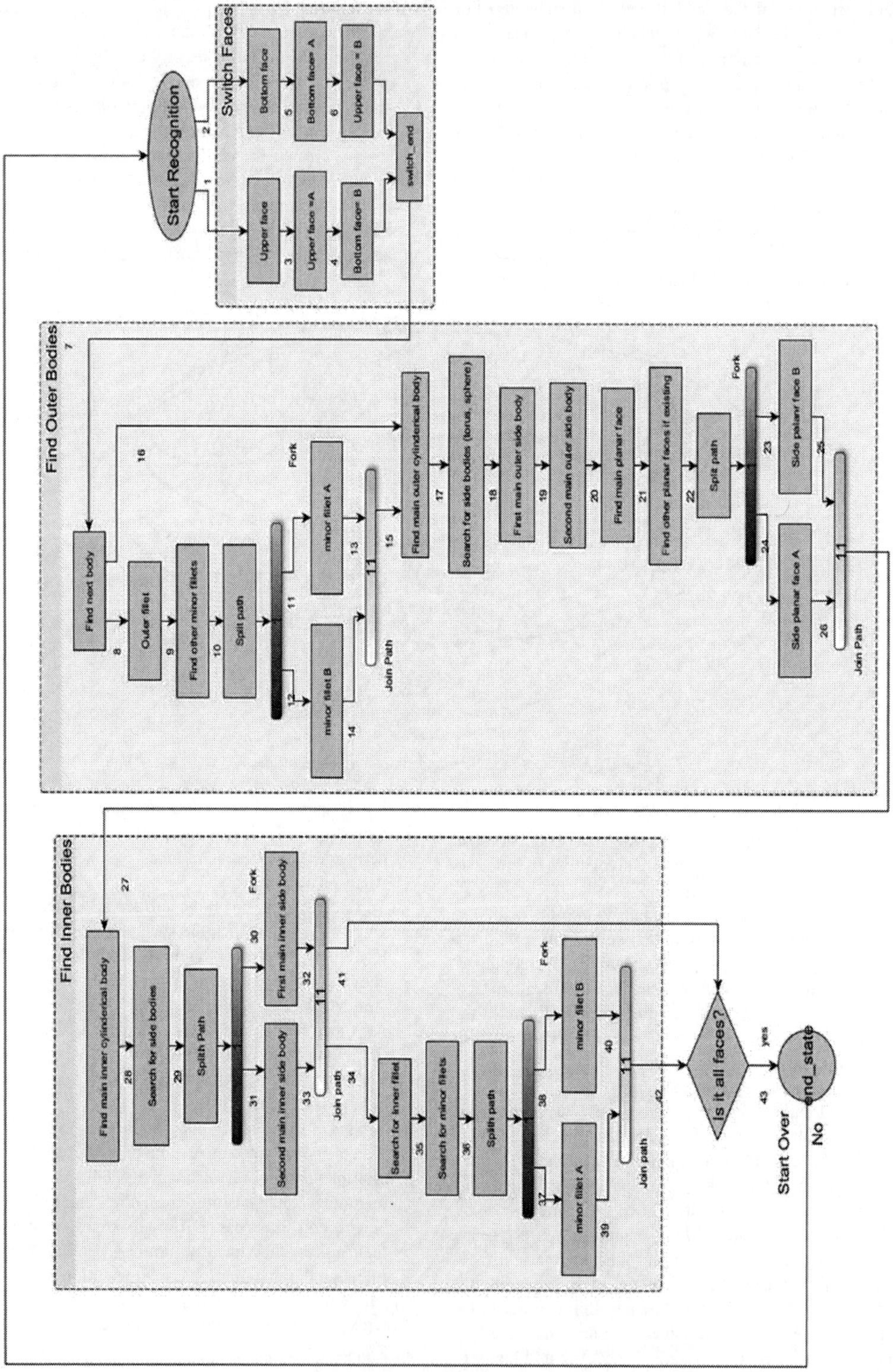

Fig. 7 Louver feature recognition algorithm

5. Technical details: In Fig. 8, the system writes down the found features and all possible combinations that might occur during the recognition phase. The system could make all required comparisons with the defined models, and already tried different paths for recognition from other different features, and successfully it could recognize the louver feature within 492 ms.

```
</FREngine::Start>
<Results n_end_states=1 duration=492ms />
<FREngine::Start>
<Transition name='1' path info='' n out='1'>
<Transition name='2' path info='' n out='1'>
<Transition name='5' path info='->2' n out='1'>
<Transition name='6' path info='->2' n out='1'>
<Transition name='join switch' path info='->2' n out='1'>
<Transition name='7' path info='->2->join switch' n out='1'>
<Transition name='8' path info='->2->join switch->7' n out='0'>
<Transition name='16' path info='->2->join switch->7' n out='0'>
<Transition name='3' path info='->1' n out='1'>
<Transition name='4' path info='->1' n out='1'>
<Transition name='join switch' path info='->1' n out='1'>
<Transition name='7' path info='->1->join switch' n out='1'>
<Transition name='8' path info='->1->join switch->7' n out='1'>
<Transition name='16' path info='->1->join switch->7' n out='1'>
<Transition name='17' path info='->1->join switch->7->16' n out='1'>
<Transition name='18' path info='->1->join switch->7->16->17' n out='0'>
<Transition name='9' path info='->1->join switch->7->8' n out='1'>
<Transition name='10' path info='->1->join switch->7->8->9' n out='1'>
<Transition name='11' path info='->1->join switch->7->8->9->10' n out='1'>
<Transition name='12' path info='->1->join switch->7->8->9->10' n out='1'>
<Transition name='14' path info='->1->join switch->7->8->9->10->12' n out='1'>
<Transition name='13' path info='->1->join switch->7->8->9->10->11' n out='1'>
<Transition name='15' path info='' n out='1'>
<Transition name='17' path info='->15' n out='1'>
<Transition name='18' path info='->15->17' n out='2'>
<Transition name='19' path info='->15->17->18(2)' n out='1'>
<Transition name='20' path info='->15->17->18(2)->19' n out='1'>
<Transition name='21' path info='->15->17->18(2)->19->20' n out='1'>
<Transition name='22' path info='->15->17->18(2)->19->20->21' n out='1'>
<Transition name='23' path info='->15->17->18(2)->19->20->21->22' n out='1'>
<Transition name='24' path info='->15->17->18(2)->19->20->21->22' n out='1'>
<Transition name='26' path info='->15->17->18(2)->19->20->21->22->24' n out='1'>
<Transition name='25' path info='->15->17->18(2)->19->20->21->22->23' n out='1'>
<Transition name='27' path info='' n out='1'>
<Transition name='28' path info='->27' n out='1'>
<Transition name='29' path info='->27->28' n out='1'>
<Transition name='30' path info='->27->28->29' n out='1'>
<Transition name='31' path info='->27->28->29' n out='1'>
<Transition name='33' path info='->27->28->29->31' n out='1'>
<Transition name='32' path info='->27->28->29->30' n out='1'>
<Transition name='34' path info='' n out='1'>
<Transition name='41' path info='' n out='0'>
<Transition name='35' path info='->34' n out='1'>
<Transition name='36' path info='->34->35' n out='1'>
<Transition name='37' path info='->34->35->36' n out='1'>
<Transition name='38' path info='->34->35->36' n out='1'>
<Transition name='40' path info='->34->35->36->38' n out='1'>
<Transition name='39' path info='->34->35->36->37' n out='1'>
<Transition name='42' path info='' n out='1'>
<Transition name='43' path info='->42' n out='1'>
<EndState name='LouverFeatureEndState'/>
<Transition name='19' path info='->15->17->18' n out='0'>
</FREngine::Start>
<Results n_end_states=1 duration=1012ms />
```

Fig. 8 All possible transitions, and found faces during the feature recognition process

It is worth mentioning that these tests have been done only for testing, and with the final solution's release version speed is expected to increase between 20 and 100 times.

IX. Conclusions

In this paper we introduced our current work with feature recognition for sheet metal parts. Here, we have developed a system for feature identification, done the required integration, designed recognition algorithms and gone through multiple phases of testing and optimization. At first, the presented system does interface with 3D CAD and CAM tools, and a high level of programming was done to ensure correct recognition without any flaws. For precision, we considered the different sub-features that could exist within the same feature, and we could handle that by performing a thorough classification to all features and grouping them into different patterns. Following that, we combined the different variations with their similar algorithms, so in some cases one algorithm might handle up to 512 sub-features of the main feature. Finally, we performed intensive testing with either ideal models, or real parts coming from customers. Currently the system can handle up to 11 different features which are the most common ones to exist, with recognition successfulness up to 100%. The recognition system at a later stage does interface with the tooling sub-system that takes all processed data about features, their locations and their properties, so that it can specify the tools required to perform a certain model or a feature within a model. In the current phase we are trying to include more features even if they are practically minor features that might not occur or by the best very rarely, and we are aiming at this phase to introduce a complete system that can handle all sheet metal features.

References

[1] Nnaji, Bartholomew O., et al. "Feature reasoning for sheet metal components." The International Journal of Production Research 29.9 (1991): 1867-1896.

[2] Farsi, Mohammad Ali, and Behrooz Arezoo. "Feature recognition and design advisory system for sheet metal components." Proceeding of the 5th IATS (2009).

[3] Huhtala, M., M. Lohtander, and J. Varis. "Product data management and sheet metal features—Sheet metal part recognition for an easier designing process producing manufacture-friendly products." 2013 IEEE International Conference on Industrial Engineering and Engineering Management. IEEE, 2013.

[4] Groover, Mikell, and E. W. J. R. Zimmers. CAD/CAM: computer-aided design and manufacturing. Pearson Education, 1983.

[5] Mercer, Tim. CAD/CAM selection for small manufacturing companies. Diss. University of Wisconsin, 2000.

[6] Justin, Mitchell. "Types of CAD Software." Techwalla. N.p., n.d. Web. 21 Dec. 2016. <https://www.techwalla.com/articles/types-of-cad-software>.

[7] Siemens, P. L. M. "Software Inc." NX Nastran User's Guide (2008).

[8] Cannon, Larissa, et al. "How can NX Advanced Simulation support multi-user design?" Comput.-Aided Des. Appl., PACE (2), Aug (2012): 21-32.

[9] FRE is a copyright and a registered property of Prima Power, Metallitie 4, FI-62200, Kauhava, Finland.

[10] Kannan, T. R., and M. S. Shunmugam. "Processing of 3D sheet metal components in STEP AP-203 format. Part I: feature recognition system." International Journal of Production Research 47.4 (2009): 941-964.

[11] Kannan, T. R., and M. S. Shunmugam. "Processing of 3D sheet metal components in STEP AP-203 format. Part II: feature reasoning system." International journal of production research 47.5 (2009): 1287-1308.

[12] Fournier, Sue. Sheet Metal Handbook. Vol. 575. Penguin, 1989.

[13] Nigam, Swami D., and Joshua U. Turner. "Review of statistical approaches to tolerance analysis." Computer-Aided Design 27.1 (1995): 6-15.

Investigation of Green Dye-Sensitized Solar Cells Based on Natural Dyes

M. Hosseinnezhad, K. Gharanjig

***Abstract*—**Natural dyes, extracted from black carrot and bramble, were utilized as photosensitizers to prepare dye-sensitized solar cells (DSSCs). Spectrophotometric studies of the natural dyes in solution and on a titanium dioxide substrate were carried out in order to assess changes in the status of the dyes. The results show that the bathochromic shift is seen on the photo-electrode substrate. The chemical binding of the natural dyes at the surface photo-electrode were increased by the chelating effect of the Ti(IV) ions. The cyclic voltammetry results showed that all extracts are suitable to be performed in DSSCs. Finally, photochemical performance and stability of DSSCs based on natural dyes were studied. The DSSCs sensitized by black carrot extract have been reported to achieve up to Jsc=1.17 $mAcm^{-2}$, Voc= 0.55 V, FF= 0.52, η=0.34%, whereas Bramble extract can obtain up to Jsc=2.24 $mAcm^{-2}$, Voc= 0.54 V, FF= 0.57, η=0.71%. The power conversion efficiency was obtained from the mixed dyes in DSSCs. The power conversion efficiency of dye-sensitized solar cells using mixed Black carrot and Bramble dye is the average of the their efficiency in single DSSCs.

***Keywords*—**Anthocyanin, dye-sensitized solar cells, green energy, optical materials.

I. Introduction

DSSCs have more and more attention due to low cost, vast range of materials, transparency and green technology. The photosensitizers have direct effect on photovoltaic properties of DSSCs [1], [2]. The natural dyes, due to low cost, abundant and no environmental treat, are appropriate for using in DSSCs device. Specially, natural dyes have an importance role in green DSSCs [3]. Anthocyanins are derivatives of antocyanidins that, depending on their pH, may appear red, purple or blue. Anthocyanins occur in leaves, stems, roots, flowers and fruits. The bonding between natural dye and photoanode substrate has been formed by carbonyl and hydroxyl groups in anthocyanin molecules. The excited electron of dye molecules has been injected through this bonding [4]. Polo et al. assembled DSSCs cells using two fruits from anthocyanin groups. The results show that the power conversion efficiency of these fruits is 1.5% and 0.95%, respectively [5]. Nishanta et al. used *Kopsia flavida* fruit as natural photosensitizer in DSSCs and achieved η=0.38% [6]. Four natural photosensitizers based on natural photosensitizers have been investigated in dye-sensitized solar cells by Kushwaha et al. and they achieved power conversion efficiency of 0.12%-0.37% [7]. Wongcharee et al. assembled DSSCs based on rosella, blue pea and mixed extracts. The highest power conversion efficiency of 0.37% has been obtained for rosella extract. The results show that the temperature of extraction processing decreased the performance of devices [8]. Fernando et al. fabricated DSSCs cells based on natural dyes extracted from tropical flowers. The highest power conversion efficiency of 1.14% has been obtained for hibiscus flowers [9]. Bathi et al. studied the performance of DSSCs based on *Lawsonia inermis* leaves, sumac/rhus fruits and *Curcuma longa* roots. The highest power conversion efficiency of 1.5% has been obtained for red purple sumac [10]. Ozuomba et al. studied the optical and photovoltaic performance of the DSSCs based on Hibiscus flower. The natural photosensitizer was the member of anthocyanin family. The power conversion efficiency of sensitized DSSCs was 0.58% [11]. Park et al. extracted natural photosensitizer from gardenia yellow and used in DSSCs structure. The maximum absorption and power conversion efficiency of natural dyes were 544 nm and 0.17%, respectively [12].

In this study, two natural dyes, Black carrot and Bramble, have been used as photosensitizers. The UV-Vis spectra of natural dye in solution and on photo-anode substrate were studied and absorption maxima wavelength and intensities of the extraction were also investigated. Finally, these natural extraction and mixture have been used in DSSCs and their photovoltaic behaviors have been studied.

II. Experimental

A. Materials and Instruments

All chemical materials and solvents utilized in this study were analytical grade provided by Merck Co. without further purification. The sample of Bramble and Black carrot were collected from natural source in Boshehr city of Iran during 2016 growing seasons. Transparent conducting oxide, FTO (F-doped SnO_2, DyeSol), TiO_2 pastes, scattering layer (Sharif Solar Co.) were purchased. The UV-Vis spectrophotometry and FTIR spectra were measured with Cecil 9200 double beam transmission spectrophotometer and Perkin-Elmer Lamada 25, respectively.

B. Sample Preparation

Brambles and Black carrot are members of the anthocyanin group that are appropriate for using in dye-sensitized solar cells structure. Anthocyanin is polar pigment with carbonyl and hydroxyl groups (Fig. 1) that may appear in various colors from pink. These groups can be linked to the photo-anode

M. Hosseinnejad is with the Institute for Color Science and Technology, Tehran, Iran (corresponding author to provide Phone: 98-21-22944184; Fax: 98-21-22947537; e-Mail: hosseinnezhad-mo@icrc.ac.ir)

K. Gharanjig is with the Institute for Color Science and Technology, Tehran, Iran and Center of Excellence for Color Science and Technology, Tehran, Iran (e-mail: gharanjig@icrc.ac.ir) surface in DSSCs structure. The general chemical structure of anthocyanin and picture of the investigated natural dye source are shown in Fig. 1. In this study, the sample of Black carrot and Bramble was collected from underbrush grown in south of Iran. The extract of Black carrot and Bramble were achieved from fresh fruit and their performance is investigated in dye-sensitized solar cells.

C. Electrochemical Measurements

Electrochemical measurements of the synthesized dyes were carried out in solution in acetonitrile. The oxidation potential (E_{ox}) was measured using three small-sized electrodes. A quasi reference electrode (QRE) was used as the reference. Platinum wires were used as the working and the counter electrodes. All electrode potentials were calibrated with respect to ferrocene (Fc)/ferrocenium (Fc^+) redox couplet. An acetonitrile solution (2 ml) of dyes containing tetrabutylammonium perchlorate (0.1 mol dm^{-3}) and ferrocene (ca. 1 mmol dm^{-3}) was prepared. The electrochemical measurements were performed at a scan rate of 100 mV s^{-1} [13].

Fig. 1 (a) Bramble, (b) black carrot and (c) general structure of anthocyanin

D. DSSCs Assembly

A nanocrystalline TiO_2 film was coated on a FTO glass glass support. The dye solutions were adsorbed by dipping the coated glass for overnight in a solution of the dye in ethanol. Finally, the film was washed with an ethanol solvent. The electrolyte was Acetonitrile-ethylenecarbonate (v/v=1:4) containing tetrabutyl ammonium iodide (0.5 mol dm^{-3}). The components of DSSCs contain photoanode; Pt counter electrode and electrolyte solution were prepared in sandwich method [14]. Monochromatic light was used to investigate the action spectrum. Illumination with AM 1.5 was utilized for investigating photocurrent-photovoltage curves by a Bunko-Keiki CEP-2000 system.

III. Results and Discussion

Anthocyanins are the water soluble pigment of natural dyes that adsorb light at the longest wavelength. Anthocyanin molecules have carbonyl and hydroxyl substituents as functional groups that can be linked to the surface of a porous titanium dioxide substrate. This bonding transfers the excited electron of dye molecules to the conduction band (CB) of metal oxide such as TiO_2. The KBr plates must be thoroughly cleaned after this procedure to prevent contamination of future samples. The infrared (FT-IR) spectra were studied and the results demonstrated appearance of important peak. The OH group (bonding vibration) has appeared in 3331 cm^{-1} and 3361 cm^{-1} which demonstrated that hydrocyle groups of natural extract contain anthocyanin that causes bonding between metal oxide and functional groups, respectively. The board band centered at 500-600 cm^{-1} is due to the vibration of the Ti-O band in the porous nano anatase titanium dioxide substrate [15], [16].

Anthocyanin compounds exhibit a wide band in UV-Vis region of the spectrum due to charge transfer transition. Table I illustrated the wavelength of maximum absorption (λ_{max}) and the molar absorpivity (ε_{max}) for the Black carrot and Bramble and UV-Vis absorption spectra for natural dyes shown is in Fig. 2, together with the λ_{max} of the corresponding dyes adsorbed on photo-anode substrate. Upon dye adsorption onto a titanium dioxide surface, the wavelength of maximum absorption is shifted by 10 and 13 nm (bathochromic shift) for black carrot and bramble extract, respectively as compared to the corresponding spectra in solution. Their binding can be increased by the chelating effect to the Ti(IV) ions [5]. The attachment to the titanium dioxide surface confirms the exited state, thus shifts toward the lower energy of the absorption maximum [15], [17]. Table I shows the molar extinction coefficients of Black carrot and Bramble extract in solution at their respective λ_{max}, indicating that these natural dyes have good light harvesting abilities [6].

The cyclic voltammetry method has been used for studying oxidation potential (E_{ox}) of natural dye. The cyclic voltammetry spectra illustrate two redox waves that first oxidation wave shows oxidation of the internal standard of ferrocene and second wave indicates the electrochemical oxidation of the dye. Cyclic voltammetry method is an appropriate experimental way for study of oxidation potential of natural dye [6]. The oxidation peak potential (E_{pa}) for bramble and black carrot as natural dye can therefore be calculated as 0.66 vs and 0.69 vs Fc/Fc^+ in solution.

TABLE I
Absorption of the Natural Dyes

Dye	λ_{max} (nm) (in solution)	ε ($M^{-1}cm^{-1}$)	λmax (nm) (on TiO_2)
Black carrot	489	27254	491
Bramble	521	27579	541

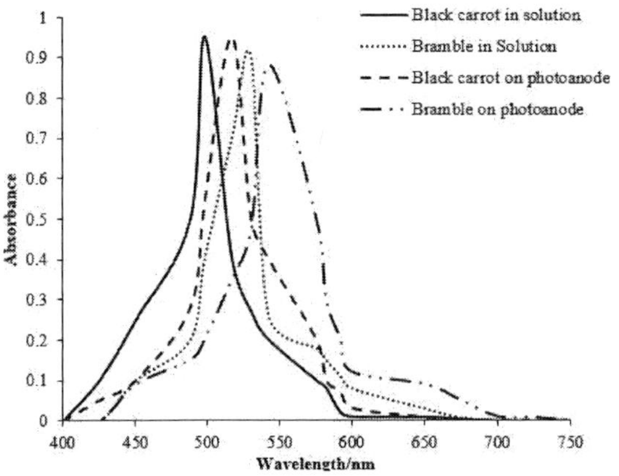

Fig. 2 UV-Vis absorption spectra for natural dyes

No reduction peak was observed for the synthesized dye. Therefore, the E_{ox}-$E_{0\text{-}0}$ level, where $E_{0\text{-}0}$ represents the intersection of normalized absorption and the fluorescence spectra in solution (Fig. 2), was calculated. This is considered to correspond to the reduction potential [16], [18]. The $E_{0\text{-}0}$ of organic dye was observed corresponding to 2.01 eV and 2.11 eV, respectively. Therefore, the E_{ox}-$E_{0\text{-}0}$ level of organic dye is calculated to be –1.33 V and -1.29 V vs Fc/Fc$^+$ in acetonitrile.

We extract anthocyanin from Black carrot and Bramble for making natural photosensitizer for dye-sensitized solar cells. The photocurrent–photovoltage (J–V) diagrams of devices sensitized by natural dyes have been shown in Fig. 3. Photovoltaic parameters of DSSCs based on natural dyes have been presented on Table II.

TABLE II
PHOTOVOLTAIC PERFORMANCE OF DSSCS BASED ON BLACK CARROT AND BRAMBLE

Dye source	J_{SC} (mA.cm^{-2})	V_{OC} (V)	FF (%)	η (%)
Black carrot	1.17	0.55	0.52	0.34
Bramble	2.24	0.54	0.57	0.71

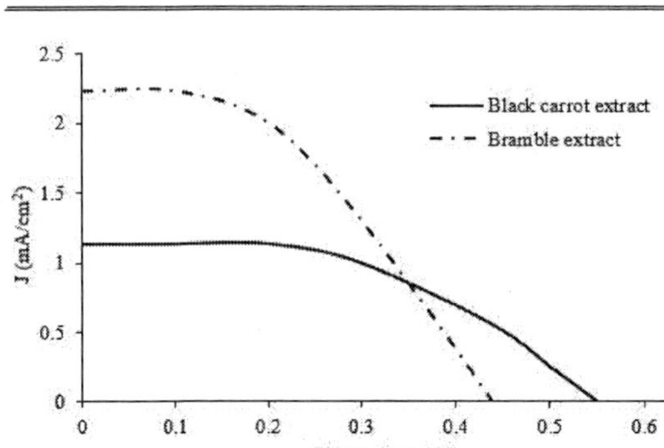

Fig. 3 Current density-voltage characteristics for black carrot and bramble

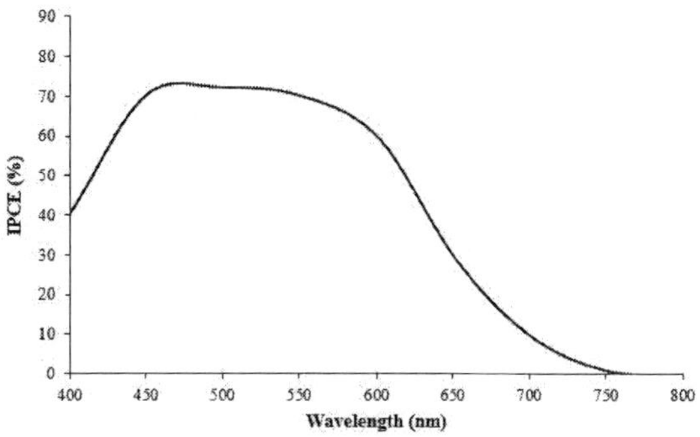

Fig. 4 IPCE spectra of DSSCs based on black carrot extraction

According to the results shown in Table II, the power conversion efficiencies of DSSCs based on Black carrot and Bramble extraction under AM 1.5 are 0.34% and 0.71%, respectively. Bramble extract shows better charge transfer due to higher light absorption; thus, Bramble is better photosensitizer compared to Black carrot [9], [10]. Thus, under similar assembling processing of DSSCs and investigation conditions, power conversion efficiency of Bramble extract is higher than Black carrot.

The photo to current conversion efficiency (IPCE) spectra of DSSC based on black carrot extraction is presented in Fig. 4. The natural dye shows IPCE ranged from 75%. Finally, we fabricated dye-sensitized solar cells using mixer of both natural pigments extracted from Bramble and Black carrot. The short circuit photocurrent (J_{sc}), open-circuit voltage (V_{oc}), fill factor (FF) and conversion efficiency (η) of DSSC based on mixture extraction are 1.64 mA cm^{-2}, 0.52 V, 0.55% and 0.47%, respectively. The dye-sensitized solar cell sensitized by a mixed Bramble and Black carrot shows a conversion efficiency of 0.47%.

IV. CONCLUSIONS

Two natural dyes native to Iran were collected and extracted from Bramble and black carrot. The natural dyes, due to low cost, abundant and no environmental treat, are appropriate for using in DSSCs device. The UV-Vis of natural extraction in solution and on photo-anode substrate was investigated. The results show that the absorption maxima wavelength of Black carrot and Bramble are 489 nm and 541 nm, respectively. The photo-anode sensitized by Black carrot and Bramble extraction shows bathochromic phenomena (32 nm and 50 nm) compared to the dye spectra in solution. Finally, dye-sensitized solar cells are synthesized using anthocyanin extracted from Bramble and Black carrot. The power conversion efficiencies of DSSCs based on Black carrot and Bramble are 0.34% and 071%, respectively. The DSSCs have been fabricated using mixed extract. The power conversion efficiency of these devices is 047% (average value of single DSSCs based on two natural dyes). The results show that the Bramble illustrates the best power conversion efficiency due to appropriate interaction between the functional groups of natural dye and photo-anode substrate.

ACKNOWLEDGMENT

The authors sincerely thank the Center of Excellence for Color Science and Technology for making this investigation possible.

REFERENCES

[1] M. Hosseinnezhad, S. Moradian, K. Gharanjig, "Investigation of effect of anti-aggregation agent on the performance of nanostructure dye-sensitized solar cells", *Opto-electron. Rev.*, vol. 23, pp. 126-130, 2015.
[2] M. Hosseinnezhad, S. Moradian, K. Gharanjig, F. Afshar Taromi, "Synthesis and Characterization of Eight Organic Dyes for Dye-Sensitized Solar Cells", *Mater. Technol.*, vol. 29, pp. 112-117, 2014.
[3] D.D. Babu, D. Elsherbing, H. Cheema, A. El-Shafei, A.V. Adhikari, "Highly efficient panchromatic dye-sensitized solar cells: synergistic interaction of ruthenium sensitizer with novel co-sensitizers carrying different acceptor units", *Dyes Pigment*, vol. 132, pp. 316-328, 2016.
[4] S. Hao, J. Wu, Y. Huang, J. Lin, "Natural dyes as photosensitizers for dye-sensitized solar cells", *Solar Energy*, vol. 80, pp. 209-214, 2006.
[5] A. S. Polo, N. Y. Tha, "Blue sensitization for solar cells natural dyes from Calafate and Jaboticaba", *Sol. Energy Mater. Sol. Cells*, vol. 90, pp. 233-238, 2006.
[6] M. Nishanta, Y. Yapa, V. Perera, "Sensitization of electrochemical solar cells with a natural dye extracted from Kopsia Flavida fruit", *Proceed Tech. Sess.*, vol. 28, pp. 54-58, 2012.
[7] R. Kushwaha, P. Sirvastava, L. Bahadur, "Natural pigments from plants used as sensitizers for TiO_2 based dye-sensitized solar cells", *Journal Energy*, vol.11, pp. 653953-654961, 2013.
[8] K. Wongcharee, V. Meeyoo, S. Charadej, "Dye-sensitized solar cells using natural dyes extracted from rosella and blue pea flowers", *Sol. Energy Mater. Sol. Cell*, vol. 91, pp. 566-571, 2007.
[9] J.M.R.C. Fernado, G.K.R. Sandeera, "Natural anthocyanins as photosensitizers for dye-sensitized solar cells", *Curr. Sci.*, vol. 95, pp. 663-665, 2008.
[10] S.A.M. Bathi, I. Alaei, I. Sopyan, "Natural photo-sensitizers for dye-sensitized solar cells", *Int. J. Renew. Energy Res.*, vol. 3, pp. 138-143, 2013.
[11] J.O. Ozuomba, L.U. Okali, A. J. Ekpunobi, "The performance and stability of anothocyanin local dye as a photosensitizer for DSSCs", *Adv. Appl. Sci. Res.*, vol. 4, pp. 60-69, 2013.
[12] K.H. Park, T.Y. Kim, S. Han, H.S. Ko, S.H. Lee, Y. M. Song, J.H. Kim, J. W. Lee, "Light harvesting for dye-sensitized solar cells", *Spectrochim Acta Part A: Mol. Biomol Spect.*, vol. 128, pp. 868-873, 2014.
[13] J. Chang, C. Lee, D. Kumar, P. Chen, L. Lin, K.R. Thomas, K. Ho, "Co-sensitization promoted light harvesting for dye-sensitized solar cells using unsymmetrical squaraine dye and novel pyrenoimidazole-based dye", *J. Power Sources.*, vol. 240, pp. 779-785, 2013.
[14] M. Matsui, T. Fujita, Y. Kubota, K. Funabiki, The use of indoline dyes in a zinc oxide dye-sensitized solar cell, J. Jin, T. Yoshida, H. Miura, *Dyes Pigments*, vol. 80, pp. 233-238, 2009.
[15] C. G. Garcia, A. S. Polo, N. Y. Murakami, "Fruit extracts and ruthenium polypyridinic dyes for sensitization of TiO_2 in photoelectrochemical solar cells", *J. Photochem. Photobiol. A: Chem.* Vol. 160, pp. 87-91, 2003.
[16] N. A. Ludin, A. M. Mahmoud, A. B. Mohamad, A. A. Kadhum, K. Sopian, N. Abdul Karim, "Review on the development of natural dye photosensitizer for dye-sensitized solar cells", *Renew. Sust. Energy Rev.*, vol. 31, pp. 386–396, 2014.
[17] H. Yang, Y. Yen, Y. Hsu, H. Chou, and J. T. Lin, "Organic Dyes Incorporating the Dithieno(3,2-b:2′,3′-d)thiophene Moiety for Efficient Dye-Sensitized Solar Cells", *Org. Lett.*, Vol. 12, No. 1, pp. 16-19, 2010.
[18] H. Hosseinnezhad, R. Jafari, K. Gharanjig, "Characterization of a green and environmentally friendly sensitizer for a low cost dye-sensitized solar cells", *Opto-Electron Rev.*, vol. 25, pp. 93-98, 2017.

M. Hosseinnejad was born in 1984 in Tehran, Iran, and received hers B.SC. from K.N. Toosi University of Technology in 2008 and hers M.SC. from the Amirkabir university of technology in 2010. After this, she studied and received his Ph.D. from Amirkabir University of Technology in 2014 under the direction of Professor Siamak Moradian, on a thesis entitled "the synthesis of organic dyes based on indoline for dye-sensitized solar cells". After hers doctoral degree, she was appointed as an assistant professor at the Institute for Color Science and Technology in 20014. Hers current research interests include the design and synthesis of organic dyes and dye sensitizers for dye-sensitized solar cells.

AUTHOR INDEX